34 00/
2

D1174964

THE MILITARY

More Than Just a Job?

Brassey's Titles of Related Interest

Art REORGANIZING AMERICA'S DEFENSE: LEADERSHIP IN WAR AND PEACE

Bowman THE ALL-VOLUNTEER FORCE AFTER A DECADE

Deitz GARRISON: TEN BRITISH MILITARY TOWNS

Jolly CROWN PROPERTY?: THE MILITARY MAN, HIS FAMILY AND THEIR COMMUNITY

Liddle HOME FIRES & FOREIGN FIELDS: BRITISH SOCIAL AND MILITARY EXPERIENCE IN THE FIRST WORLD WAR

Segal LIFE IN THE RANK AND FILE: ENLISTED MEN AND WOMEN IN THE ARMED FORCES OF THE U.S., AUSTRALIA, CANADA, AND THE U.K.

Sweetman SWORD AND MACE: TWENTIETH CENTURY CIVIL-MILITARY RELATIONS IN BRITAIN

Related Journal

(Free specimen copy available upon request)

Defense Analysis

THE MILITARY

More Than Just a Job?

Edited by

Charles C. Moskos & Frank R. Wood

PERGAMON-BRASSEY'S
International Defense Publishers, Inc.

Washington · New York · London · Oxford
Beijing · Frankfurt · São Paulo · Sydney · Tokyo · Toronto

U.S.A. (Editorial)	Pergamon-Brassey's International Defense Publishers, 8000 Westpark Drive, Fourth Floor, McLean, Virginia 22102, U.S.A.
(Orders)	Pergamon Press, Inc., Maxwell House, Fairview Park, Elmsford, New York 10523, U.S.A.
U.K. (Editorial)	Brassey's Defence Publishers Ltd., 24 Gray's Inn Road, London WC1X 8HR
(Orders)	Brassey's Defence Publishers Ltd., Headington Hill Hall, Oxford OX3 0BW, England
PEOPLE'S REPUBLIC OF CHINA	Pergamon Press, Room 4037, Qianmen Hotel, Beijing, People's Republic of China
FEDERAL REPUBLIC OF GERMANY	Pergamon Press GmbH, Hammerweg 6, D-6242 Kronberg, Federal Republic of Germany
BRAZIL	Pergamon Editora Ltda, Rua Eça de Queiros, 346, CEP 04011, Paraiso, São Paulo, Brazil
AUSTRALIA	Pergamon-Brassey's Defence Publishers Pty. Ltd., P.O. Box 544, Potts Point, N.S.W. 2011, Australia
JAPAN	Pergamon Press, 5th Floor, Matsuoka Central Building, 1-7-1 Nishishinjuku, Shinjuku-ku, Tokyo 160, Japan
CANADA	Pergamon Press Canada Ltd., Suite No. 271, 253 College Street, Toronto, Ontario, Canada M5T 1R5

Copyright © 1988 Pergamon-Brassey's International Defense Publishers, Inc.

First edition 1988

Library of Congress Cataloging-in-Publication Data
The military - more than just a job?
Includes bibliographies.
1. Sociology, Military. 2. Armed Forces—Vocational guidance. I. Moskos, Charles C. II. Wood, Frank R.
U21.5.M58 1988   306'.27   87-32850

British Library Cataloguing in Publication Data
The military: more than just a job?
1. American military forces. Personnel.
Sociological perspectives
I. Moskos, Charles C. II. Wood, Frank R.
306'.27'0973

ISBN 0-08-034321-X

Printed in Great Britain by A. Wheaton & Co. Ltd., Exeter

Contents

Part Two. Issues in the American Military

Part Three. Comparative Perspectives

Part Four. Institution versus Occupation Reconsidered

Contents

List of Figures

List of Tables

Acknowledgments

This book represents the contributions of 17 authors who have done research on and thought about the social organization of the military. Some are civilians, others are military officers, and all are exceptionally qualified scholars of armed forces and society. The genesis of this volume goes back to June 1985, when the International Conference on Institutional and Occupational Trends in Military Organization was held at the U.S. Air Force Academy. In attendance were social scientists from eight Western countries in addition to the United States. The purpose of the conference, supported by the Air Force Office of Scientific Research, was to appraise the utility of the institution-to-occupation thesis for broader understanding of trends in military organization.

This volume, in part, is a result of that conference. After the conference, most of the participants reworked their papers to be included here. For the sake of thematic unity, we found it necessary to exclude some of the papers presented at the conference, but several additional chapters were written especially for this publication. The original conference papers are found in the appendix to "Institutional and Occupational Trends in Military Organization," final technical report to the Air Force Office of Scientific Research (AFOSR F49620-84-C-0061: 1986).

We extend our thanks to General David C. Jones (U.S. Air Force, retired), former chairman of the Joint Chiefs of Staff and chief of staff of the air force, who was the first military policy maker to recognize the value of the institution-versus-occupation formulation and who sponsored much of the early social scientific work in this area. Our thanks also to those farsighted Air Force officers who helped provide the organizational support for the international conference Lieutenant General Kenneth L. Peek, Major General Robert C. Oaks, Colonel Robert C. Roehrkasse, Colonel Peter L. Conklin, Lieutenant Colonel Freeman L. Johnson, and Lieutenant Colonel Roger W. Alford. Although this book presents a rethinking of the conference proceedings, these military professionals must be credited for planting the seeds of the scholarly thought presented in these pages. We would also like to thank the following conference participants, whose active involvement enriched our comprehension of armed forces and society: Robert Gest, Franklin D. Margiotta, David Marlowe, David R. Segal, H. Wallace Sinaiko, and Thomas C. Wyatt.

Elizabeth Pereyra Kurubas not only efficiently typed a complex manuscript, but also was able to make sense of copy given her in bits and pieces and never in order. Charles C. Moskos would also like to acknowledge the Woodrow Wilson International Center for Scholars and the U.S. Army Research Institute for the Behavioral and Social Sciences, whose support at different times gave him the opportunity to develop the central concepts that underlie the research reported in this volume.

Finally, the usual caveat that the authors alone are responsible for the findings and interpretations presented is especially relevant here. Indeed, because the findings are based on materials collected from nine countries, there is probably something in here to upset some defense officials in each one of those countries. We hope so.

They looked upon themselves as men who lived by higher standards of behavior than civilians, as men who were the bearers and protectors of the most important values of American life, who maintained a sense of discipline while civilians abandoned themselves to hedonism, who maintained a sense of honor while civilians lived by opportunism and greed.

from *The Right Stuff*

I go anywhere in the world they tell me to go, any time they tell me to, to fight anybody they want me to fight. I move my family anywhere they tell me to move, on a day's notice, and live in whatever quarters they assign me. I work whenever they tell me to work . . . I don't belong to a union and I don't strike if I don't like what they're doing to me. And I like it. Maybe that's the difference.

from *A Country Such as This*

Part One

Institution versus Occupation

I

Introduction

CHARLES C. MOSKOS AND FRANK R. WOOD

Discussion of the future of the armed forces usually involves a concern with technological developments or global strategy. Most members of the armed forces, however, understand and experience the military as a social organization. This is a book about the social organization of the armed forces. It is meant to be conceptual, factual, and practical: conceptual in that it presents a paradigm of organizational change in the military, factual because it draws on case studies that advance while modifying the original paradigm, and practical in the sense that it sets forth principles on which to base organizational and personnel policies.

The military maintains its autonomy but also refracts societal trends. Because of this duality, two models of organization—institutional and occupational—describe alternative conceptions of the military. The overarching hypothesis is that the American military has been moving away from an institutional format to one that increasingly resembles that of an occupation.

When the institutional/occupational (I/O) thesis was first proposed in 1977, it gained attention not only from social researchers on military organization but also from a number of senior commanders in the armed forces. The military was undergoing a strange sort of crisis: not the noisy turbulence of the latter years of the Vietnam War but a quiet malaise, a sense that the recruits were being bought at the margin of the labor market, that officers were driven by careerism, and that reasons for military service had become obscured. The heart of the matter, it seems, was a paralysis in the definition of the essence of the military organization.[1] It is also noteworthy that "institutional versus occupational" terminology was entering the political vocabulary in discussions of military compensation and the all-volunteer force (AVF).[2]

To describe the move toward occupationalism is not to hold that such a trend is inevitable. Indeed, the very recognition of the trend focused attention on the consequences of policies that affect military social organization, and led to renewal of institutional thought and initiatives in the armed forces. The I/O thesis is to be understood as a warning, not a prediction: a caution, not an indictment.

Like other intellectual developments, the I/O thesis was not received uniformly by all groups. It conflicted with the econometric mindset that had become dominant in the Office of the Secretary of Defense and in sponsored research on military manpower.[3] A consequence of econometric analysis is to downplay the less tangible noneconomic factors and value-driven aspects of military organization. Econometric analysts prefer typically to deal with the material dimensions because they are the only ones that can be measured easily. The econometric model is a little too neat, however; it ignores the fact that the armed forces are not merely fluid collections of self-maximizing individuals, but sets of social relations and institutional arrangements as well.

The I/O thesis provided the main counterbalance to the econometric mindset in military manpower policy. Indeed, one writer has characterized the military manpower debate as "economics" versus "sociology."[4] Although this view oversimplifies the contending sides, the dichotomy has a kernel of truth. The econometric approach tends to define issues that are amenable to existing methodologies and thus concentrates on narrowly conceived comparisons of variables to the neglect of the more difficult issues of institutional change and civilian-military relations.

The I/O thesis recast the terms of the military manpower debate by introducing a component previously neglected by leading civilian policy makers and analysts. The real contribution of the I/O thesis, a sociological construct in contrast to the econometrics of the policy analyst, is the explicit recognition of the trade-offs involved in military manpower choices. The I/O thesis reintroduced a consideration of contextual factors. In place of the individual atomism of the econometric mold, it presented military life in a dynamic organizational framework.

In recent decades, the members of the armed forces have felt increasing conflict between internal pressures toward institutional integration and societal trends that push toward identification with like occupational groups in the larger society. The argument about the proper relationship between the service member's internal ties within the military and his external links with those outside the military—between institution and occupation—will never reach a clear-cut conclusion. Most of the time members of the military will prefer something of each. The pressing question is this: does a tilt toward civilianization make any real difference in military effectiveness? We think that it does. The results of creeping occupationalism can be found in three key areas: mission performance, member motivation, and professional responsibility. Let us look briefly at each of these areas.

From the standpoint of mission performance, the basic assumption (well grounded in research) is that institutional identification fosters greater organizational commitment and performance than does occupational.[5] The armed forces require certain behavior from their members that can never be made to serve individual interests, certainly not in a narrow economic sense. Internalization of institutional values implies nearly unbounded definitions of tasks and the manner in which these tasks are to be carried out. The logic of occupationalism, conversely is to define task boundaries and to set standards of accomplishment that, if met, signify adequate performance. In general, an occupation pays enough to fill the job and to get it done—no more. An effective manager in an occupation prevails on workers to do what they are supposed to do; an effective leader in an institution motivates members to do more than they are supposed to do.

A second effect of occupationalism is to mask organizational developments that replace the intrinsic motivation of an institution with the extrinsic motivation of an occupation. A large body of social psychological research documents the difference between intrinsic motivation, as in action due to personal values, and extrinsic motivation, as in behavior brought about by pay.[6] The interaction of intrinsic and extrinsic rewards can be quite complex; not only may these rewards be nonadditive, but also inducing members to perform tasks with strong extrinsic rewards may create behavior that will not be performed in the future except for even greater extrinsic rewards. Extrinsic rewards, moreover, can weaken intrinsic motivation. As an example, let us imagine a military organization that decides to "motivate" its members in a crucial skill position by paying them extra for that skill. In this situation, we might expect that increasing the extrinsic reward will increase the effort expended. However, the military person's motivation may possibly decrease sharply if the extrinsic reward is no longer forthcoming, and the member's interest in performing military activities outside the specified task *without extra pay* may be reduced correspondingly.

A third consequence of occupationalism is somewhat more insidious—the undermining of military professionalism. If military functions can be reduced to dollars, then ultimate decisions on the military organization and military personnel become the province of cost-benefit analysts; decisions are removed from the military profession. An institutional approach, on the contrary, never loses sight of the uniqueness of military organization in a democratic society. The nation has entrusted its armed forces with responsibilities rarely, if ever, found in civilian life: defending the national interest, the real possibility that military members will risk life and limb in that role, and, in recent years, the awesome responsibility of deploying and guarding the nuclear arsenal.

This is the policy context in which the social scientific utility of the I/O thesis emerged. In the decade following the initial presentation of this thesis, more than 30 studies have been completed that either seek to test the I/O thesis in some statistical manner or deal with the I/O thesis in a conceptual fashion. Questionnaire items seeking to tap I/O dimensions are provided in the appendix; a selected bibliography of I/O research is also given. Although the original I/O formulation was based on the particulars of the American case, much subsequent research has been conducted in Western military systems outside the United States. Significantly, from a research standpoint, trends originally specified in the United States were found to be occurring in other NATO countries—and perhaps beyond. The time had come to assess, update, and codify the I/O thesis.

The plan of the book is straightforward. The first part sets forth the conceptual and societal-level effects of the I/O thesis. Part two applies the I/O thesis to specific areas in the American military. The third part is a comparative examination of I/O trends in eight countries, each with its own history and civil-military format. The final part comes full circle with an appraisal of the I/O thesis for military personnel policy in the United States.

INSTITUTION VERSUS OCCUPATION

Part one provides an analysis of issues in the development and applicability of the I/O thesis. Despite some differences in methods and emphases, the authors' conclusions are consistent; they all recognize that the I/O thesis refers to a number of levels of analysis. On one level, we are speaking about shifts in individual attitudes and behavior. At another level, we are discussing changes within the social organization of the armed forces. At a third level, we are interested in how broad historical and societal trends affect civil-military relations. To appreciate the explanatory power of the I/O thesis, the researcher must understand how these levels interact, even if the researcher focuses only on a single level.

Charles C. Moskos sets the stage by presenting an updated version of the original I/O thesis, which retains the basic distinction between the value orientation of the institution and the rational calculation of the marketplace. Differences between institution and occupation are specified along such variables as societal regard, role commitments, reference groups, basis and mode of compensation, legal system, reference groups, residence, sex roles, and evaluation of performance. Significantly, Moskos notes that the marked trend toward occupationalism in the 1970s has been countered somewhat by a renewed emphasis on institutionalism in the military of the 1980s.

Frank R.Wood sees the air force officer corps as the military group most susceptible to occupational trends; thus, his detailed examination of air force officers illustrates the ascendancy of technology and civilian comparability within a military organization. Wood conceives of the I/O thesis as having different salience at the micro and the macro levels. At the macro level, military decisions are shared increasingly with civilians; advancing technology tends to fragment the institutional center and to aggravate the reliance on civilians. Most important, the core function of the air force—flying—loses its centripetal significance in the organization. At the micro level, military members, especially pilots and navigators, identify increasingly with civilian counterparts.[7] For support officers, the identification with managers is most apparent. The overall trend is away from those who identify mainly as military officers and toward those who see themselves as specialists in uniform, a movement from military professionals to professionals in the military.

Charles A. Cotton, whose earlier work on the Canadian forces was one of the most important empirical studies of the I/O thesis, presents a broad theoretical and programmatic overview of the thesis.[8] He argues that unless steps to the contrary are taken, societal change normally causes military organizational change, which in turn causes individual change. Cotton believes that an isolated military in a democratic society is impractical as well as undesirable. At the same time, "economic man," or a military based on marketplace principles, can lead to the worst of two worlds: a military isolated from civilian society and a lack of cohesion within the military. The optimum is a military organization that is both internally cohesive and congruent with civilian society. To accomplish this goal, military leaders must articulate institutional values and emphasize consistently the organizational priority of operational readiness. Otherwise, societal pressures toward occupationalism will continue to be ascendant. Specifically, Cotton believes that military leadership consistently should shape the organizational culture to emphasize the nation over the military organization, the mission over career, the unit over the individual, and the service ethic over the job ethic.

John H. Faris makes the cardinal point that the I/O dichotomy exists basically between institutional values and bureaucratic rationalism. Bureaucratic rationalism contains an implicit understanding of motivation of a type that is concerned primarily with objective calculation. Bureaucratic rationalism sets expectations that are specific, stable, and predictable. The problem is that the styles of leadership consistent with bureaucratic rationalism do not take nonrational and moral concerns into account. Institutional values foster a leadership style that seeks to achieve more than "the expected" by unleashing energies, perhaps affected strongly by personal emotions, that are congruent with a vision of the organization's goals. In a sense, the I/O thesis is another way of stating the conflict between idealism and materialism, between romantics and utilitarians. Faris also notes that despite the dominance of marketplace thought in the policy elite of all-volunteer forces, institutional values continue to operate at important levels within the armed forces themselves. From a methodological viewpoint, Faris holds that reliance on survey methods is itself a form of bureaucratic rationalism, but that if we need surveys we should probe the attitudes of senior civilian policy makers rather than those of service members.

ISSUES IN THE AMERICAN MILITARY

Part two deals with "social issues" confronting the American armed forces. The purpose of this section was to select familiar personnel topics and to see how they could be illuminated by the I/O thesis. This procedure cast new light on old issues and also led to certain policy recommendations that sometimes deviated from the conventional wisdom. In addition, applying the I/O thesis to concrete topics facilitated modifications in the thesis and allowed for more productive formulations; these were tested, so to speak, against "real world" problems.

Mady Wechsler Segal maintains that the military and the family are greedy institutions; both demand strong commitment from their members. Thus, career members of the military are in a special type of double bind. The military institution has traditionally been demanding because of the risks undertaken by its members, the frequency of family relocations and separations, and residence in foreign countries. The military, however, has not established new normative patterns to accommodate contemporary trends: more married males at the junior enlisted level, more female service members, more single parents in the military, the appearance of dual-service couples, and the entrance of military spouses into the civilian labor force. The net result is that the military is more in conflict with the family than ever before. Segal suggests that, paradoxically enough, the military may have to become less greedy of its family members in order to maintain its own institutional qualities.

One important outcome of the AVF has been the increasing reliance of the armed services on female personnel. Patricia M. Shields points to the continuing, and perhaps insurmountable, obstacles in the way of full incorporation of women into the military. She argues that the military has both unique attractions and particular drawbacks for women. Equal pay for equal rank is a powerful attraction, but the pressures of parenthood impinge much more heavily on military women than on military men, and more on military women than on civilian women. The question is whether increasing numbers of women will alter the prevailing male ethos of the armed forces; ultimately some accommodation must be made between women's parental responsibilities and military demands. Without such an accommodation, the full incorporation of women into the military remains problematic.

John Sibley Butler examines race relations from the vantage of the I/O thesis. His starting point is the original assumption of the I/O thesis, namely, that societal and organizational changes affect individual behavior. It follows that as the military becomes more occupational, race relations will resemble increasingly those in civilian life. The unique success of the military in race relations inheres largely in its institutional quality, which downplays premilitary social status and fosters equal contact between the races. Yet several occupational trends augur unfavorably for race relations in the military: white enlisted members tend more than blacks to live off base, thus changing the unit's racial composition at night; certain military occupational specialties tend to become defined as predominantly black or white; the civilian stratification system is replicated in the race/class makeup of the contract civilian force. The upshot of Butler's analysis is that the "spillover" of any racial strife in the society at large into the military will be greater to the degree that armed forces are occupational.

The military academies might be regarded as the extreme case of institutional socialization in the armed forces. Although inculcating institutional values is a prime function of the academies, William H. Clover and Thomas M. McCloy of the U.S. Air Force Academy staff find that I/O conflicts are not absent from the service academies. On the one hand, the institutional side of academy life is fostered by emphases on character, athletics, and spiritual concerns; on the other, occupational tendencies can be found in the academic curriculum, which converges increasingly with civilian institutions of higher education and marketing strategies that portray the academy along those lines. Survey data of cadets show that overall cadets are very high on institutional values, but that this orientation recedes somewhat the longer they remain at the academy. In addition, women cadets are more likely than men to reflect an institutional orientation.

COMPARATIVE PERSPECTIVES

Part three introduces a cross-national aspect to the I/O thesis. Although the thesis was derived originally from the particulars of the American experience, the concept quickly attracted the attention of social researchers of military systems outside the United States. Because of this international interest, contributors to this volume include leading social scientists representing eight Western countries: Great Britain, France, the Federal Republic of Germany, Australia, the Netherlands, Greece, Switzerland, and Israel.

A comparative focus on issues of military organization is rare in the social research on armed forces. Even rarer is the imposition of a single conceptual scheme—in this case, the I/O thesis—on a set of cross-national studies. Comparative analyses present pitfalls and opportunities. The main pitfall is that the missions confronting the United States are much more extensive than those faced by any other Western military force, and the problems of the American military are more complicated than those encountered by most other countries. Even so, comparative research offers an opportunity to demonstrate how core military social functions can be performed by different kinds of structures. Although always wary of false analogies, comparative analysis makes possible a much more powerful understanding of the social trends in the American military.

Historically, the British forces have been among the most institutional in the Western world. As Cathy Downes points out, the British forces remain largely institutional despite certain trends toward occupationalism in the current period. Since the beginning of modern times, British forces have been much more marginal to British society than have American forces in the United States. An important qualification, however, is the connection between the military and the hereditary elites, notably the royal family. In 1982, Prince Andrew was a combatant in the Falklands War. (However, four years later, an uproar developed in the United Kingdom over Prince Edward's decision *not* to complete his tour of service with the Royal Marines.) For centuries, British forces have been concentrated in colonial service, sea duty, and isolated home garrisons. In the contemporary British military, recruitment advertising is pitched heavily toward institutional appeals, with strong emphasis on self-sacrifice and serving the common good. Although the British forces are all volunteer, they have undergone little pressure to adjust to changing societal norms. Relative to the United States, promotion criteria are subjective and decentralized. British forces, however, have partially adopted pay comparability with civilian occupations, and they offer specialty pay to retain needed technicians. Private home ownership is widespread, moreover. Finally, in recent years the wives of career officers have played a relatively minor role in military community activities.

Bernhard Fleckenstein describes the military of the Federal Republic of Germany as occupational in a special sense. With the inauguration of the Bundeswehr in the 1950s, a conscious effort was made to create a civically integrated military; the prototype was the civil service. Thus, the military in West Germany allows unions (for career members), is moving to adopt a form of overtime pay, and relies on civilian courts for punishment of service members. Furthermore, recruitment appeals are almost purely occupational, with emphasis on compensation and technical training. Noncommissioned officers, especially those in technical branches, are the most occupational component in the German military. Yet, despite these features, significant latent institutional qualities exist in the Bundeswehr; indeed, if a discernible trend exists, it is toward reinstitutionalism. Senior officers are generalists and try to counter tendencies toward technical specialization in the officer corps. Officer leaders are strongly evident in the supervision of soldiers and, in a manner of speaking, bypass noncoms in the guidance of enlisted personnel. Conscription continues to be accepted in West Germany, which also tends to foster an institutional orientation.

As in West Germany, the French military is a mixture of draftees and career members, but as described by Bernard Boëne, the I/O configuration of the French forces differs from that of the Bundeswehr in major ways. Among Western forces, the French remain relatively institutional; much of the compensation is in kind rather than in cash. Differences exist among services; however, the air force is most occupational, the army most institutional, and the navy somewhere in between. As a rule of thumb, the more technical the branch, the more occupational it seems to be. In general, the career force lives off base, and there are signs of growing resentment of military interference in the service member's private and family life. Much of the prestige of the noncommissioned officer corps comes not from military status but from the simple fact that a senior noncom is well paid. Boëne makes a major contribution to the sociology of the military by presenting a chart of I/O tendencies by rank and branch.

Nicholas A. Jans's study of Australia portrays an all-volunteer military with a strong institutional residue, moving toward occupationalism. A salary system is replacing traditional forms of military compensation; the notion of 24-hour commitment is weakening; extra-organizational grievance channels have replaced the traditional chain of command. Almost all the military/family strains noted in the American forces are repeated in the Australian case. In particular, officers are reluctant to relocate when their wives have good employment. Survey data show that Australian Air Force pilots have a strong identity as specialists rather than as military officers. Jans notes that occupational types feel committed when they perceive their compensation as good, but become extremely uncommitted when military compensation falls behind civilian rates. On the other hand, institutionalists become disaffected in the face of what they perceive as ascendant occupationalism. Notably, Jans cites survey data showing that erosion of service benefits is seen as resulting from ineffective military leadership, which in turn has been accompanied by an upsurge of support for trade unionism in the Australian military.

Jan S. van der Meulen gives an account of the armed forces of the Netherlands. Despite the Dutch armed forces' reputation for being a military of unions, unkempt soldiers, and working to rule, van der Meulen argues that a strong institutional undertow cannot be discounted. To begin with, the Dutch nation and its draftees accept the legitimacy of conscription. Although noncoms have strong occupational orientations, they continue to recognize the distinctiveness of military life. The career officers show almost a conscious resistance to becoming military managers, and (as we saw in the case of the Bundeswehr) a movement toward reinstitutionalization can be detected. In the Dutch military, the animosity of the career officer corps seems directed especially toward meddling politicians, pacifists, feminists, and sociologists. Van der Meulen concludes that the Dutch forces are occupational on the outside and institutional on the inside.

The Greek military seems to be diametrically opposed to the Dutch, as Dimitrios Smokovitis takes for his benchmark the Greek military of the 1950s through the 1970s. During this period, the Greek military was almost purely institutional, archaically so: the soldier's basic identity was with the military, not with any specialty; draftees could not wear civilian clothes even off duty; promotion of officers to senior ranks was limited to generalists; salary was entirely a function of rank and seniority; pay was nominal for draftees and not much better for junior officers; the status of career military officers' wives corresponded to their husbands' ranks; all crimes (including traffic offenses) were judged by military court-martial. Although the Greek military still remains heavily institutional, occupational signs are evident in the contemporary period: competitive salaries have been introduced to recruit military technicians; a growing number of officers attend universities with concomitant nonmilitary identities; officers' wives are entering the work force increasingly; there has been some retrenchment in the purview of military justice. In the Greek case, the trend toward occupationalism corresponds with a lesser likelihood that the military will intervene in the civil political order.

Switzerland is the prime example of the militia system; Karl Haltiner describes the Swiss military system as institutional but nonprofessional. The Swiss force is based on universal service (for males) and is extremely value-oriented. Because there is no professional officer corps or cadre, militia officers and noncoms, often members of civilian elites, perform many extra hours of duty for which they are essentially uncompensated. Yet even in Switzerland, Haltiner notes occupational trends; advancing weapon technology requires specialization and civilianization of some military functions; elite youth show a notable decline in willingness to serve beyond the minimum requirement, with resulting difficulties in attracting high-quality noncoms and officers.

The final case study concerns Israel, perhaps the most institutional military in the Western world. Reuven Gal points out that the Israeli Defense Force (IDF) is almost completely value-driven; the society hardly questions the need for an effective military. The IDF consists of a permanent service force, which receives decent compensation but takes early retirement; conscriptees, who serve with virtually no salary; and reservists. Unlike other military systems, the Israeli reserve forces are the core of the military, not an appendage. The military role is seen as generalist rather than specialist, diffuse rather than specific. Veteran status is virtually a prerequisite for achievement in civilian society. Underlying changes in the contemporary period include more administrative centralization, less contact between leaders and subordinates, more public scrutiny of armed force, some internal questioning of involvement in aggressive wars, and an incipient trend among some officers to view the military in occupational rather than service terms. Nevertheless, the IDF presents itself as an almost ideal institution, clear about its mission in the nation's defense.

A comparative overview of the I/O thesis allows for the following generalizations. Institutional militaries can be either fully integrated with civilian society, as in Israel and Switzerland, or somewhat isolated, as in the case of the United Kingdom. The West German and Dutch military systems have an occupational gloss, but are characterized by a strong institutional undertow in the career force. The French military, and to some extent the Greek, are almost the opposite: a dominant institutionalism confronts an occupational undertow. The Australian forces seem to be changing rapidly from institutional to occupational. In comparison, U.S. forces are closer to the occupational end of the spectrum than most comparable militaries, but with some movement toward reinstitutionalization.

INSTITUTION VERSUS OCCUPATION RECONSIDERED

The final chapter is a summary perspective that seeks to uncover the core principles of institutionalism. We repeat the importance of understanding the I/O thesis at micro, organizational, and macro levels for policy implications.[9] Three basic conditions of institutionalism are assessed: acceptance of hardship by those in charge, clear vision of what the organization is all about and how the parts relate to the whole, and the awareness that institutional members are value-driven. In addition, institutional features can occur in a variety of organizational forms, service members who do not display traditional institutional traits should not be "read out" of the institution, and the trend toward specialization and advanced technology fragments institutional identity only when members of the military begin to identify with their civilian counterparts.

The military need not act in haste to curtail all moves toward occupationalism; the system can absorb a certain amount of occupationalism, and if we consider the matter carefully, creative measures can be taken to counter the drawbacks. In this vein, we make several broad recommendations for American military manpower policy, with special attention to recruitment and retention, the family, sex roles, organizational commitment, and leadership. Adopting measures to restore institutionalism in the American military does not mean turning back the clock. Rather, it means striking a new balance after the indiscriminate acceptance of many occupational features.

NOTES

1. A partial list of critical assessments of the American military just since 1980 makes up a library in itself. See James Fallows, *National Defense* (New York: Harper and Row, 1982); Richard A. Gabriel, *To Serve with Honor* (Westport, Conn.: Greenwood, 1982); Edward R. Luttwak, *The Pentagon and the Art of War* (New York: Simon and Schuster, 1984); Richard A. Gabriel, *Military Incompetence* (New York: Hill and Wang, 1985); James Coates and Michael Kilian, *Heavy Losses* (New York: Viking, 1985); Gary Hart with William S. Lind, *America Can Win* (Bethesda, Md.: Adler and Adler, 1986); Arthur T. Hadley, *The Straw Giant* (New York: Random House, 1986); and Richard Halloran, *To Arm a Nation* (New York: Macmillan, 1986).

2. *Military Compensation: Key Concepts and Issues* (Washington, D.C.: General Accounting Office, Jan. 10, 1986) presents an explicit discussion of institutional versus occupational concepts. See also Robert L. Goldich, "Military Nondisability Retirement 'Reform,' 1969–1979," *Armed Forces & Society* 10, no. 1 (Fall 1983): 59–85.

3. See, for example, Richard. V.L. Cooper, *Military Manpower and the All-Volunteer Force* (Santa Monica, Calif.: Rand, 1977); *America's Volunteers*, a report isued by the office of the Assistant Secretary of Defense (Manpower, Reserve Affairs and Logistics), December 31, 1978; and *Military Manpower Task Force*, a report to the president on the status and prospects of the all-volunteer force, November 1982.

4. Robert H. Baldwin, Jr., "The Decision to End the Draft: Economics vs. Sociology," *Joint Perspectives* 3, no. 1 (Summer, 1982): 58–68.

5. On normative commitment and organizational effectiveness, see Richard T. Mowday, Lyman W. Porter, and Richard M. Steers, *Employee-Organization Linkages* (New York: Academic Press, 1982).
   Of special relevance are those studies that deal directly with normative determinants and military

cohesion. In this regard, see for both content and methodological innovation John H. Johns, ed., *Cohesion in the Military* (Washington, D.C.; National Defense University Press, 1984); and William Darryl Henderson, *Cohesion* (Washington, D.C.: National Defense University Press, 1985). Both these military studies were strongly influenced by the I/O thesis. Also relevant is Gwyn Harries-Jenkins and Jacques van Doorn, eds., *The Military and the Problem of Legitimacy* (Beverly Hills, Calif.: Sage, 1976); Gwyn Harries-Jenkins, ed., *Armed Forces and the Welfare Societies* (London: Macmillan, 1982); and Gwyn Harries-Jenkins and Charles C. Moskos, "Armed Forces and Society," *Current Sociology* 29, no. 3 (Winter 1981): 1–170.

6. Barry M. Staw, *Intrinsic and Extrinsic Motivation* (Morristown, N.J.: General Learning Press, 1976), especially pp. 169–185.

7. Attitudes and values of American pilots are reported in Frank R. Wood, "Air Force Junior Officers: Changing Prestige and Civilianization," *Armed Forces & Society* 6 (Spring 1980): 483–506; and James H. Slagel, "The Junior Officer of the 1980s," *Air University Review* 33 (November–December 1981): 90–96. For Australia, See Nicholas A. Jans, "Institutional and Occupational Orientations in the Australian Defence Force," paper presented at the International Conference on Institutional and Occupational Trends in Military Organization, U.S. Air Force Academy, Colorado, June, 1985. For Canada, see John F. Bennett, "Professional Attitudes of Canadian Forces Junior Officers," Research Report, Air War College, Maxwell Air Force Base, Ala., March 1985.

8. Charles A. Cotton, "Institutional and Occupational Values in Canada's Army," *Armed Forces and Society* 8 (Fall 1981): 99–110.

9. The classic treatments of military professionalism also simultaneously dealt with micro, organizational, and macro levels. See Samuel P. Huntington, *The Soldier and the State* (Cambridge, Mass.: Harvard University Press, 1957); Morris Janowitz, *The Professional Soldier* (New York: Free Press, 1960).

II

Institutional and Occupational Trends in Armed Forces

CHARLES C. MOSKOS

The argument is that the armed forces of the United States are moving from an organizational format that is predominantly institutional to one that is becoming more and more occupational. The contrast between institution and occupation is easy to overdraw, of course. To characterize the armed forces as either an institution or an occupation is to do an injustice to reality. Both elements have been and always will be present in the military system. But the social analyst must always use pure types to advance conceptual understanding. Our concern is to grasp the whole, to place the salient fact, and to have a framework to appraise the relevant policy. Even though terms like *institution* or *occupation* have descriptive limitations, they do contain core connotations that serve to distinguish each from the other.

Although the discussion that follows draws heavily from the American experience, the essential differences between institutional and occupational (I/O) models of military organization are phrased in terms suitable for cross-national research. These differences are summarized in Table 2-1. The I/O thesis assumes a continuum ranging from a military organization highly divergent from civilian society to one highly convergent with civilian structures.

Concretely, of course, military forces have never been entirely separate or entirely coterminous with civilian society, but the conception of a scale, along which the military more or less overlaps with civilian society, highlights the ever-changing interface between the armed forces and society. This also alerts us to emergent trends within the military organization. Over the years, incremental developments slowly amount to profound changes. A shift in the rationale of the military toward the occupational model implies organizational consequences in the structure and, perhaps, the function of armed forces.

TABLE 2-1. Military Social Organization: Institutional vs. Occupational

Variable	Institutional	Occupational
Legitimacy	Normative values	Marketplace economy
Societal regard	Esteem based on notions of service	Prestige based on level of compensation
Role commitments	Diffuse; generalist	Specific; specialist
Reference groups	"Vertical" within the armed forces	"Horizontal" with occupations outside the military
Recruitment appeals	Character qualities; life-style orientation	High recruit pay; technical training
Evaluation of performance	Holistic and qualitative	Segmented and quantitative
Basis of compensation	Rank and seniority; decompressed by rank	Skill level and manpower shortages; compressed by rank
Mode of compensation	Much in noncash form or deferred	Salary and bonuses
Legal system	Military justice; broad purview over member	Civilian jurisprudence; limited purview over member
Female roles	Limited employment; restricted career pattern	Wide employment; open career pattern
Spouse	Integral part of military community	Removed from military community
Residence	Work and residence adjacency; military housing; relocations	Work and residence separation; civilian housing permanence
Post-service status	Veterans' benefits and preferences	Same as nonserver

An *institution* is legitimated in terms of values and norms, that is, a purpose transcending individual self-interest in favor of a presumed higher good. We use *institution* here in the sense it usually possesses in everyday speech. Members of an institution are often seen as following a calling captured in words like *duty, honor,* and *country.* They are commonly viewed and regard themselves as being different or apart from the broader society. To the degree institutional membership is congruent with notions of self-sacrifice and primary identification with one's institutional role, institution members ordinarily enjoy esteem from the larger society.

Military service traditionally has acquired many institutional features, for example, fixed terms of enlistment, liability for 24-hour service, frequent moves of self and family, subjection to military discipline and law, and inability to resign, strike, or negotiate working conditions. When grievances are felt, members of an institution do not as a rule organize themselves into interest groups. Rather, if redress is sought, it takes the form of personal recourse to superiors, with its implication that the organization will take care of its own. Above and beyond these conditions, of course, there are the physical dangers inherent in combat training and actual combat operations.

Moreover, a paternalistic remuneration system, corresponding to an institutional model, evolved in the military: much of compensation is noncash ("in kind")—such as food, housing, uniforms, and medical care; subsidized base consumer facilities; payment to service members partly determined by family size; and a large proportion as deferred pay in the form of retirement benefits. To the degree military service is based on the citizen-soldier concept, pay for recruits is below the market wage, although there may be postservice benefits. Notions of overtime pay are alien to the institutional military. In addition, unlike many civilian compensation systems in which marketability determines reward, remuneration in the military is essentially based on rank and seniority.

An *occupation* is legitimated in terms of the marketplace. Supply and demand, rather than normative considerations, are paramount. Workers with equivalent skill levels ought to receive approximately the same pay, whatever the employing organization. In a modern industrial society, employees usually enjoy some voice in the determination of appropriate salary and work conditions. Such rights are counterbalanced by responsibilities to meet contractual obligations. The cash-work nexus emphasizes a negotiation between individual (or workers' groups) and organizational needs. A common form of interest articulation is the trade union. The occupational model implies the priority of self-interest rather than that of the employing organization.

The occupational military model is anchored in marketplace principles. Whether under the rubric of econometrics or that of systems analysis, such redefinition of the military is based on a set of core assumptions: (1) no analytical distinction exists between the military and other systems, in particular, no difference between cost-effectiveness analysis of civilian enterprises and military services; (2) military compensation should as much as possible be in cash, rather than in kind or deferred, thereby allowing for a more efficient operation of the marketplace; and (3) military compensation should be linked directly to skill differences of individual service members.

Beyond military sociology, polarities of social structures have been at the core of mainstream macrosociology. One need only mention Max Weber on traditional versus legal-rational authority, Henry Maine on status versus contract, Emile Durkheim on organic versus mechanical solidarity, Ferdinand Toennies on *Gemeinschaft* versus *Gesellschaft*, and Talcott Parsons on the shift from collective to individualistic pattern variables. Any master trend in Western society would certainly be found somewhere along the lines of the shift from normative to functional integration. The posited shift of the military system from institutional to occupational is perhaps no more than the particular application to the military of this master trend.

INSTITUTIONAL AND OCCUPATIONAL MILITARIES COMPARED

Traditionally, the American military has sought to avoid the organizational outcomes of the occupational model. There has been a reluctance to adopt a salary system that would incorporate all pay, allowances, and benefits into a single cash salary, despite the recommendations of numerous governmental bodies to adopt such a system. Nevertheless, even the traditional military system has made some accommodation to occupational imperatives. Bonuses or off-scale pay have long been incentives used to retain physicians, certain other professionals, and expensively trained personnel in technical skills.

Despite certain exceptions, the conventional system of military compensation reflects the corporate whole of military life. The military institution is organized *vertically*, whereas an occupation is organized *horizontally*. People in an occupation tend to feel a sense of identity with others who do the same sort of work and receive similar pay. Horizontal identification implies key reference groups are external to the organization. In an institution, on the other hand, it is the conditions under which people live and work that develop the sense of identity that binds them together. The organization one belongs to creates the feeling of shared interest, not the work performed. In the armed forces, the very fact of being part of the same organization has traditionally been more salient than the fact that military members do different jobs.

Role commitment in an institutional military tends to be diffuse; members are expected to perform tasks not limited to their military specialities. Members are under the purview of the military organization whether on or off duty, whether on or off base. In an occupational military, role commitments tend to be job specific. The organization is not concerned with the worker's behavior away from work if it does not affect job performance.

In an institutional military, work and residence locales are adjacent. Members typically reside in military housing. Frequent relocations are understood to be part of military life. The on-base military club is often a center of social life. An occupational military has much more separation of work and residence locales. Members often live off base in rented or owned civilian housing. Permanence of residence becomes a value. Recreational social life takes place off base.

In a manner of speaking, the role of institutional membership in the military community extends to spouses (until very recently, almost always meaning wives). They are expected to initiate and take part in a panoply of social functions and volunteer activities in the military community. Military families are supportive of, or adjunct to, organizational purpose. In the occupational military, however, wives at both noncommissioned and junior officer levels are increasingly reluctant to take part in customary social functions. (The growing numbers of women in the American military have produced growing numbers of military husbands, a category virtually unacknowledged, much less researched.) With a rising proportion of wives employed outside the home, moreover, fewer wives have either the time or the inclination to engage in the volunteer work that underlies much of the social life of military installations.

In a traditional military, women service members are small in number and assigned to limited support roles, often in separate female corps. Career patterns are prescribed and restricted. In an occupational military, both recruitment needs and greater entry of women in the labor force lead to a higher proportion of female service members. Female corps are abolished, and women are much more integrated into mainstream roles. Combat exclusion strictures, however, still work against completely open career patterns. Accordingly, pressures to do away with female combat exclusion become stronger.

An institutional military tends to evaluate its personnel according to "whole person" criteria, to rely heavily on qualitative and subjective evaluations, and to favor decentralized promotion systems. An occupational military tends toward judgments relating to specific performance standards, prefers numerical or quantitative evaluations, and favors centralized promotion. The more institutional a military, the wider the span of the military justice system; the more occupational the military, the more likely it is that offenders will be tried by civilian courts. In a society characterized by an institutional military, prior military status carries over into civilian life; veterans will enjoy preferences over nonveterans, especially in government employment and entitlements.

TRENDS AND COUNTERTRENDS

Although antecedents predate the appearance of the all-volunteer force (AVF), the end of the draft might be seen as a major thrust to move the military toward the occupational model. The selective service system was premised on the notion of citizen obligation—a calling in the almost literal sense of being summoned by a local draft board—with concomitant low salaries for junior enlisted personnel. Furthermore, even volunteer recruits, many entering because of the draft, received the same low salaries as draftees. The current all-volunteer military in and of itself need not be correlated with an occupational model except that the architects of the present American AVF have chosen the occupational model as their paradigm.

The marketplace philosophy was clearly the theory underlying the 1970 *Report of the President's Commission on an All-Volunteer Force*, better known as the *Gates Commission Report*.[1] The Gates Commission was strongly influenced by laissez-faire economic thought and argued that reliance in recruiting and retaining an armed force should be primarily based on monetary inducements guided by labor force realities. The move toward making military remuneration comparable to that of the civilian sector actually preceded the advent of the AVF. Starting in 1967, military pay has been formally linked to the civil service and thus, indirectly, to the civilian labor market. During the early 1970s, military compensation, especially at recruit levels, increased at a much faster rate than civilian pay. Toward the latter part of the 1970s, military pay lagged behind civilian level. In the early

1980s, catch-up pay raises were given to military personnel, although in the mid-1980s military pay levels again appeared to be trailing civilian compensation, though the point is not without dispute.[2] What is without question, however, is the sharp rise in military pay. In 1964, the last peacetime year before the war in Vietnam, service members received an average of $15,000 a year (in 1986 dollars) in compensation; in 1986, the figure was $25,000.

Precisely because military compensation was being redefined as comparable to civilian compensation, increased attention was given to actions and proposals to reduce a number of military benefits and entitlements, most notably proposals to effect a restructuring of the retirement system. A widespread concern with "erosion of benefits" became evident among career military members since the advent of the AVF. A kind of "devil's bargain" may have been struck when military pay was geared to comparable civilian levels. Institutional features of military compensation may have been unwittingly traded off for the relatively good salaries enjoyed by military personnel in the early 1970s and again in the early 1980s. Service entitlements probably cannot be maintained at past levels if military salaries are competitive with civilian scales.

Another major outcome of the AVF has been a dramatic compression of pay scales within the military. In the 1960s, the basic pay of an E-9 (the senior enlisted grade, a sergeant major in the army) with 26 years of service was better than seven times that of an entering recruit. In 1986, that same E-9 made only four times the pay of the E-1 recruit. The ratio would have been even lower, had Defense Department recommendations prevailed in Congress. The paradox is that this front-loading of compensation toward the junior ranks and the changes to improve lower enlisted life cannot be fully appreciated by those newly entering the service because they did not experience the old ways. Second, junior enlisted members see little monetary or life-style improvement at the point at which they usually decide to enter the career force—between E-4 and E-5—thus the likelihood of their choosing to remain in the service is reduced. Once upon a time, sergeants measured their incomes and perquisites against those of their soldiers and felt rewarded; now they see a relative decline of status within the service and feel deprived when they compare their earnings against civilian wages.

The possibility that trade unions might appear within the armed forces of the United States was unthinkable two decades ago, yet reliance on marketplace models to recruit and retain military members is quite consistent with the notion of trade unions. Since 1983, moreover, members of the armed forces have been counted as part of the labor force for the first time in American history. Several unions have indicated an interest in organizing the military, but the nascent trend toward unionism was stopped by a 1978 law that prohibited any organizing activities whatsoever in the armed forces. The constitutionality of the 1978 law is yet to be tested, and the situation of full-time guardsmembers who are already unionized in state government employee unions remains unclarified.[3]

Despite the statutory prohibitions in place on organizing the armed forces, the underlying dynamics of the occupational ascendancy are still operative. A 1976 survey of air force personnel found that 33 percent of those polled stated they would join a military union, 31 percent were undecided, and 36 percent would not. Willingness to join a union was greater among enlisted members than officers and was strongly correlated with perceived erosion of benefits.[4] Another development has been the trend toward what might be called *incipient unionism* on the part of service associations. The Air Force Sergeants Association (AFSA), especially, has taken an increasingly active role in lobbying Congress for pay and benefits. Significantly, the AFSA grew from a membership of 23,000 in 1974 to over 90,000 ten years later.

In the pre-Vietnam military, enlisted personnel who did not complete their initial tours of duty were considered aberrant. During the late 1970s, however, about one in three service members were failing to complete initial enlistments and were being prematurely discharged for reasons of discipline, personality disorders, job inaptitude, and the like. (Also, about 10 percent of female service members were taking voluntary discharges for pregnancy.) Although the general recruitment picture improved tremendously in the 1980s, the attrition rate dropped only slightly. The percentages of service members from the 1982 entering cohort who had been dismissed within three years of entry were: marine corps, 32; army, 31; navy, 26; and air force, 24.

The attrition phenomenon reflects the changing policy of military separation—the easy-out system of the AVF—as well as changes in the social composition of the entering enlisted force. Put another way, the contemporary military, like industrial organizations, is witnessing the common occurrence of its members quitting or being fired. In all but name, the American military had gone a long way down the road toward indeterminate enlistments. Yet, the word *honorable*—a term not found in occupational evaluations—is still used in classifications of military discharges, despite Defense Department proposals during the 1970s to move toward general certificates of separation.

As late as the 1960s, a bachelor enlisted man living off base was practically unheard of. Not only was it against regulations, but also few could afford rent on junior enlisted pay. By the mid-1980s, although precise data are not available, a reasonable estimate would be that one of four single enlisted people in stateside bases live away from the military installation. To the increasing proportion of single enlisted members living off base, one must add the growing number of married junior enlisted people, most of whom also live in the civilian economy. Since the end of the draft, the proportion of marrieds among junior enlistees has about doubled. This trend was most noticeable in the air force; in 1986, 37 percent of male E-4s were married, as were 46 percent of female E-4s.

Another manifestation of the occupational model is the growing number of military personnel who hold outside employment. A 1979 Defense Department survey reported that close to one in five enlisted persons stated they were holding second jobs. If the data were limited to those stationed in the United States, this figure would most likely have been higher (on the presumption that second-job opportunities are less available overseas). Moonlighting is attributed to the service member's need for additional income. Yet the anomaly exists that moonlighting has increased among the junior enlisted force even though their current buying power far exceeds that of the pre-AVF era. In any event, whatever its causes, moonlighting clearly runs contrary to the institutional premise of a service member's total role commitment to the armed forces.

In recent years, military wives at both noncom and officer levels have been increasingly reluctant to initiate or take part in the customary volunteer and hosting activities long associated with military community life. With the rising number of working wives, fewer women had either the time or the inclination to engage in such activities. A 1981 air force survey showed 66 percent of enlisted wives and 45 percent of officer wives to be gainfully employed. Moreover, even those military wives who were not gainfully employed began to regauge their commitment to volunteer work in light of their perceptions of the lower effort put forth by employed wives. Female liberation among military wives, though not absent, is less important than the growing tendency for spouses to define their roles as distinct from the military community. By 1987 the refusal of spouses to quit their jobs and participate in base activities (after being pressured by some unit commanders) resulted in national media attention and the appointment of a Blue Ribbon Panel to study the issues and develop new policy.

From the 1950s through the 1960s, the federal courts, the Court of Military Appeals, and the Supreme Court brought into military law almost all of the procedural safeguards available to a civilian defendant, while narrowing the purview of military jurisdiction.[5] The highwater point in this trend was *O'Callahan* vs. *Parker* (1969), in which the Supreme Court struck down court-martial jurisdiction for non-service-connected offenses. The significance of *O'Callahan* was that the off-duty or off-base service member was to be treated like any other citizen. Within the military itself, the trend has been a shift in emphasis from courts-martial to administrative procedures, most notably in cases of early discharges. The trend toward occupationalism continued in *U.S.* vs. *Russo* (1975) and *U.S.* vs. *Larionoff* (1977), in which the Supreme Court sanctioned the applicability of basic contract law to the legal status of enlistments. The net effect of these court decisions was to move toward a legal redefinition of the military from one based on traditional status toward one more consistent with generally accepted contract principles.

The increasing proportion of civilian workers in total defense employment—from 27 percent in 1964 to 32 percent in 1985—reflects another trend in the American military establishment. The diminution of the proportion of uniformed personnel within the defense establishment appears to have slowed in the mid-1980s, but its impact on institutional commitment deserves examination.[6] Interviews and observations of military personnel working in units with civilians indicate potentially detrimental effects on morale. The narrow definition of the work role among civilians can increase the workload (such as unpaid overtime and holiday work) of military personnel. This, along with the higher pay civilians may receive for doing seemingly the same kind of work as military members, can generate resentment. The point here is that feelings of relative deprivation are unavoidable when the diffuse responsibilities of the military institution coexist with the more limited work roles found in civilian occupations.

Another manifestation of organizational change departs entirely from the formal military organization. This is the use of civilians hired on contract to perform jobs previously carried out by active-duty service members or direct-hire civilians. These tasks range from routine housekeeping and kitchen duties, through rear-echelon equipment and weapon maintenance, to ship crews of oilers and tenders and air crews of chartered aircraft, sometimes used in quasi-combat roles. Although precise figures are not available, one study showed that from 1964 to 1978, contract-hire civilians rose from 5 percent to 15 percent as a proportion of total defense staffing.[7] There is no reason to think this trend has reversed itself in the years following.

The sum of the above and related developments would seem to confirm the ascendancy of the occupational model in the emergent military, but countervailing forces are in effect. Indeed, the tension and interplay between institutional and occupational tendencies characterize organizational change within the armed forces.

Most notable have been internal military initiatives to reinvigorate institutional features in the American forces of the 1980s, a kind of counterbalance to the occupationalist ascendancy of the preceeding decade. Although the form of change has taken different directions in each of the services, the common goal has been to enhance member commitment and corporate identity. Starting in 1980, the army began moving toward a new staffing system that would replace individual replacements with whole-unit replacements in combat units. Even though the new system had been diluted to small-group replacements by the mid-1980s, the frontal recognition of the importance of unit cohesion was a trend in the institutional direction. In 1981, the navy introduced Operation Pride, whose purpose was to stress navy traditions, including more wearing of uniforms, restoring privileges of rank, and more attention to military courtesy and ceremony. In 1982, the air force launched Project Warrior, a broad-based program designed to promote service pride, awareness of the air force heritage, emphasis of leadership instead of management, and development of war-fighting awareness. Also in 1982, the Defense Department overhauled its administrative discharge program in order to restore the integrity of the honorable discharge. The new directive was "designed to strengthen the concept that military service is a calling different from any civilian occupation."[8]

At about the same time, court decisions came down that said the military and civilian worlds were separate and necessarily so. The Court of Military Appeals in 1982 in *U.S.* vs. *Lockwood* broadened off-base military jurisdiction, thus pushing *O'Callahan* in a reverse direction. The Supreme Court in 1983 in *Chappell* vs. *Wallace* stated enlisted service members could not sue their superiors for alleged violations of their constitutional rights. The high court said civilian courts must "hesitate long" before tampering with the "heart of the necessarily unique structure of the military establishment."[9]

Two Supreme Court decisions in 1987 further accentuated the separation of military from civilian law. In *U.S.* vs. *Stanley* the high court held that military personnel cannot sue the government or military superiors even for gross and deliberate violations of their constitutional rights. In *Solorio* vs. *U.S.*, the Supreme Court ruled that military personnel may be court-martialed for crimes regardless of whether the offense had any connection to military service, thus effectively repealing *O'Callahan*. One astute observer has noted, however, that although the military falls back on institutional arguments to defend command prerogatives over private behavior, such arguments may not hold if the occupational quality of the military can be effectively demonstrated. Thus, the I/O dichotomy may become a new frontier in military jurisprudence.[10]

As in all large-scale organizational change, developments in the armed forces have been shaped by several convergent and overlapping trends that are hard to disentangle from one another and whose cumulative effect is not always clear. In the 1970s, the movement toward occupationalism was prominent. In the 1980s, however, trends edged in a different direction and became more complex and contradictory. Pressures to reinstitutionalize were evident. Thus, for example, the reported number of service people working at second jobs dropped sharply; the moonlighting proportion dropped from 19 percent in 1979 to 11 percent in 1985.[11] Occupational assumptions remained predominant, however, among civilian policy makers in the Office of the Secretary of Defense.[12]

INSTITUTIONAL AND OCCUPATIONAL MODALITIES

The institutional versus occupational thesis seeks to identify an overarching trend while still recognizing that military systems are differentially shaped, depending upon a country's civil-military history, military traditions, and geopolitical position. Moreover, I/O modalities interface in different ways even within the same national military system. Differences exist between military services and between branches within these services. I/O modalities may also vary along internal distinctions, such as those between officers, noncommissioned officers, and lower ranks; between career and single-term military members; between draftees and volunteers; between active-duty servers and reservists; between technical and non-technical branches; between men and women; and between those stationed in their home countries and those stationed abroad. There may even be trends toward reinstitutionalizing the military either across the board or in specified units.

Like every theory, the I/O thesis contains an implicit understanding of motivation.[13] Is motivation rational or subjective, oriented toward moral concerns of altruism, strongly affected, perhaps, by internal emotional concerns, or is it efficient and rational, concerned primarily with objective calculation? The problematics of action are concerned with the relative weight of idealism and materialism. In philosophic terms, it is as old as the struggle between romantics and utilitarians.

Theories of change can combine rational and nonrational modes of action, and the I/O thesis offers this possibility. Because organizational changes are difficult to analyze, observers must avoid simplifying schemes that distort even as they illuminate. The I/O approach must not simply ask, What is the direction of change? as important as that question is, but also, How are these changes defined? The latter query allows us to move beyond the institutional-versus-occupational dichotomy to examine the different degrees and levels of institutional and occupational aspects and see where they are in opposition to each other and where they are manifested jointly. Such a dynamic approach comprehends not merely an either-or situation, but a shifting constellation of institutional and occupational features in the armed forces.

NOTES

1. President's Commission on an All-Volunteer Force (Gates Commission). *Report* (Washington, D.C.: Government Printing Office, 1970).
2. *Military Compensation: Selected Occupational Comparisons with Civilian Compensation* and *Military Compensation: Comparisons with Civilian Compensation and Related Issues* (Washington, D.C.: U.S. General Accounting Office, June 1986).
3. See Ezra S. Krendel and Bernard L. Samoff, eds., *Unionizing the Armed Forces* (Philadelphia: University of Pennsylvania Press, 1977); and William J. Taylor, Jr., Roger J. Arango, and Robert S. Lockwood, eds., *Military Unions* (Beverly Hills, Calif.: Sage, 1977).
4. T. Roger Manley, Charles W. McNichols, and G. C. Saul Young, "Attitudes of Active Duty U.S. Air Force Personnel Toward Military Unionization," *Armed Forces and Society* 3 (Summer 1977): 557–574.
5. James B. Jacobs, *Socio-Legal Foundations of Civil-Military Relations* (New Brunswick, N.J.: Transaction, 1986).
6. A concise treatment of civilian defense employees is in Martin Binkin with Hershel Kanter and Rolf H. Clark, *Shaping the Defense Civilian Work Force* (Washington, D.C.: The Brookings Institution, 1978).
7. Richard V. L. Cooper, *Military Manpower and the All-Volunteer Force* (Santa Monica, Calif.: Rand, 1977), p. 11.
8. *Army Times*, 15 February 1982, p. 12.
9. Air Force Times, 4 July 1983, p. 18.
10. Michael F. Noone, Jr., Columbus School of Law, Catholic University of America, personal communication, March 15, 1987.

11. *Army Times*, 29 November 1986, p. 13.
12. For a semi-official collection of articles generally favorable to the Gates Commission, see William Bowman, et al., *The All-Volunteer Force After a Decade* (McLean, Va.: Pergamon-Brassey's, 1986).
13. This formulation is adapted from Jeffrey C. Alexander, "Socio-Structural Analysis," *Sociological Quarterly* 25, no. 1 (Winter 1984): 5–26.

III

At the Cutting Edge of Institutional and Occupational Trends: The U.S. Air Force Officer Corps

FRANK R. WOOD

We tend to think of our social institutions as eternal, but they are not. They are subject to pressures for social change imposed by the societies in which they are immersed, and they must change to survive. The essential task of social scientists studying the military institution is to try to make some sense of the contingencies the military must face and to predict the changes that will result.

An effective way to track social change in an institution is to note critical changes in social organization. *Social organization* is a general term used by social scientists to describe the patterns of relationships between individuals and groups within a larger social structure. In regard to the military, the term describes the relationship of the military to the larger society, of groups within the military to each other, and of individual members to the organization in which they serve.

Recent changes in the social organization of the U.S. Air Force officer corps provide a unique example of institutional and occupational (I/O) trends in an advanced stage. Because of their extensive use of technology, the air force and the air force officer corps tend to be most susceptible to increasing specialization and a diffused sense of purpose. In short, they are at the cutting edge of I/O trends. They face the greatest pressure for occupationalism and serve as a harbinger of things to come for other branches as they become increasingly dependent on technology. Further, the air force's experience in dealing with these changes will serve as an example for other branches as they anticipate developing their own strategy to counter or adapt to these trends.

Thus, the lessons learned from the air force experience may allow others to anticipate the future and initiate change on their own terms.

This chapter will focus on social organizational changes in the military, with emphasis on how they have affected the air force officer corps. Using the air force as a case study illustrates the complex interaction of social organizational change on several levels: the macro-level changes in the relationship between the military and society, the meso-level changes in the air force as an organization, and the micro-level changes in individual attitudes toward military service and the air force. Finally, this analysis has implications for both the I/O hypothesis and air force organizational policy.

RETHINKING MODELS OF SOCIAL ORGANIZATIONAL CHANGE

In trying to assess changes in the social organization of the military, researchers have posited several models to explain the relationship of the military to society and of military members to each other. Those who study the relationship between the military and society have found it useful to place the military organization somewhere along a continuum between convergence and divergence in relation to civilian structure and norms. Those who study the officer corps in particular tend to follow the grand theorists by characterizing the officers as homogeneous or heterogeneous: how much social diversity do the officers incorporate from the larger society?

The models used by researchers in the post–World War II period reflect tremendous pressures for social change in military organization. The research that characterized the 1950s, for example, emphasized the homogeneity and divergence of the officer corps. C. Wright Mills popularized the term *military mind,* and Samuel Huntington credited the military's divergence from society as a key factor in its professionalization.[1] In the later 1950s and early 1960s, Morris Janowitz's classic work signaled a trend toward civilianization or narrowing differences between military and civilian society.[2] During the Vietnam period of the late 1960s and early 1970s, the combat function of the military became salient, and a plural model was suggested by Moskos and others.[3] In this model, military persons associated with combat activities were presumed to be divergent from society. During this period as well, the likelihood of an all-volunteer force sparked a resurgence of the divergent-homogeneous model.[4] In the mid-1970s, Moskos signaled yet another change when he proposed the institutional-occupational thesis.[5]

Although researchers appear to be confused as to the best way to view the military social organization, these models actually display a good deal of conceptual continuity. At the center of all the models is presumed to be a core of professional military officers who exalt the unique military function of combat. The models vary only in the size of this characteristic core and in the degree of convergence with or divergence from civilian structures and norms.

The I/O thesis represents an extension of these models and reflects a long-term trend toward convergence or civilianization of the military.

In the last ten years, the I/O thesis has generated much discussion and controversy. Military leaders have generally supported it because it articulates changes in social organization that they had sensed for a long time; this articulation offers hope for a solution to growing retention problems and increasing demands to "do more with less."[6] Researchers, on the other hand, suspect that the thesis is correct but are not sure how it should be conceptualized and find it difficult to measure or study scientifically. Janowitz was the first to emphasize conceptual difficulties when he criticized Moskos for "changing the rules of the game of social analysis without clearly signalling the change he introduced" and suggested that the change is better conceptualized as from professional to occupational.[7] Difficulties in conceptualization are important because they make operationalization and measurement difficult.

The I/O thesis is difficult to conceptualize and measure because it represents social organizational change on several levels as well as an interaction between these levels. On the macro level, the I/O thesis suggests that the military is moving from an institutional to an occupational organization. The institutional organization is highly divergent from civilian society and is legitimated by values, norms, and purposes that transcend individual self-interest. The occupational organization is highly convergent with civilian structures and is legitimated in terms of the marketplace.[8] Implicit in this formulation is the micro-level notion that individual orientations are also changing—from viewing one's work as a calling, or vocation, to viewing the work as just a job. Somewhere in the middle, at a meso level, the structure and operation of the military organization are also changing; work and living conditions are managed in ways typical of civilian organizations.

Conceptualizing the I/O thesis as social change on the macro, meso, and micro levels explains why systematic analysis of this trend has been problematic. Operationalization and measurement must be appropriate for the level under examination. in this regard, analysis of historical data may be more appropriate than attitude measures to explore macro changes; organizational analysis may be more appropriate at the meso level; and survey analysis or participant observation may be most appropriate at the micro level.

Past reliance on survey methods and attitudinal data to study I/O trends demonstrates an overall insensitivity to the importance of level, timing, and method, which are all critical factors. Micro-level attitudinal data is complicated by interactions with changes at other levels. Survey data is best able to provide the "rates" and "states" reported by the respondents, but it falls short when attempting to disclose the *process* that produces them. Micro-level attitudinal data, for example, may suggest that many military members no longer view the military as "a way of life," but this data cannot tell us that social and economic changes have forced spouses to work and families to live off base when they might not do so otherwise. And these changes, in turn, have produced family units anchored in the civilian community, with reduced interest in supporting military community activities or the military member's own organizational commitment.[9] Overreliance on micro-level survey data ignores the complex interaction of change on several levels: that is, social change on the macro level has affected the orientation of families on the micro level, and that change, in turn, has altered the way the military organization must operate on the meso level.

SOCIAL ORGANIZATIONAL CHANGES AT THE MACRO, MESO, AND MICRO LEVELS: THE U.S. AIR FORCE

The complex interaction of macro-level social change, meso-level organizational change, and micro-level attitudinal change is illustrated well by the case of the U.S. Air Force officer corps. On the macro level, a redefinition of the meaning of military service is signaled by a loss of professional autonomy by the military. On the meso level, this redefinition has affected organizational definitions of the core formation of the air force. On the micro level, these changes are reflected by the professional identities and commitments of military members.

Since World War II, the status of the military in society has changed, the importance of the military function has declined, and the meaning of military service is less clear than in the past. The military is no longer viewed as a special organization that performs a unique and important function critical to the survival of our society. This redefinition has been forced by at least three major social changes. First, the nature of the military task has changed. The major function now is to provide deterrence or accomplish some limited political objective. Hence, the military must share responsibility for this task with politicians and diplomats. In fact, the use of military force has become an option of last resort.[10] Second, technological change has fragmented the military organizations into many specialities and has increased reliance on nonmilitary experts (defense contractors and technical representatives) for the development and operation of weapon systems.[11] Third, in recent years limited national economic resources have caused increasing reliance on management principles and cost analysis in lieu of military expertise.[12] In short, the military has become increasingly complex and more dependent on outsiders. The issue of "who is military" and "what the military does" is no longer clear. This confusion provides the opportunity to replace military expertise and values with the more widely accepted management principles and ethics characteristic of the occupational model.

In an environment where the military function is unclear or shared with others (diplomats and technicians) and economic justification for limited resources is more important than justification based on military expertise, military professional autonomy is naturally questioned. In fact, much autonomy has been lost in recent years as civilian "watchdog" agencies have grown in number and power. Agencies like the Office of Management and Budget (OMB), the Congressional Budget Office (CBO), the General Accounting Office (GAO), and the House and Senate Armed Services Committees all have actively sought reform in the retirement system, pay and benefits, use of civilian contractors, procurement and use of weapon systems,[13] and deployment of forces. American military professionals generally subscribe to the concept of civilian control, but they expect that their military expertise will be taken into account when decisions are made. In recent years, however, the principle of civilian control has been expanded and the cost-effectiveness of the day-to-day administration of the military has been questioned. With this expansion of civilian control and the loss of professional autonomy, civilian business ethics and practices that characterize the occupational model have become prominent.

These macro-level changes are reflected at the meso level by changes in the definition of the core function of the air force organization. Core functions are important components of social organization because they focus organizational activities and structure. They also define what an organization is about. In this regard, the air force is undergoing significant change.

Over 70 percent of the officers surveyed periodically in the past ten years (n = 5,000) agree that "during [their] time in the Air Force, the prestige of the flying function has declined while that of management has increased." The highest rates of agreement tend to be found among majors and lieutenant colonels.[14] Other indications confirm this trend toward occupational structures and norms. Junior officers, for example, tend to attribute critical job characteristics of expertise, importance, and responsibility to support jobs more often than to flying jobs. Recognition from air force leadership and from civilians outside the military also tends to favor those in support specialties.[15] According to the conventional wisdom among those in operational specialties, flying officers must serve in a support or management specialty or risk not being promoted above junior officer. Indeed, most field-grade jobs are management or administrative positions, and the chances of continuing to fly in the field-grade ranks are slim. Combat experience has suffered a similar fate; in a survey of the air force elite of the 1980s (senior officers selected for the highest level of professional development), 63 percent stated that combat experience should not be a criterion for promotion to the rank of general.[16] These changes in the relative importance of core functions signal a fundamental shift in the social organization of the air force: essentially, a civilianization of what the organization is about.

Meso-level changes in the organizational definition of the core function have predictable consequences for social organization on the micro level—specifically the civilianization of professional identities and commitments of military members. Again air force officers, who are the most susceptible to occupational trends, provide a clear example of the changing social order at this level. Here the key indicators of social organization are the military members' professional identities and commitment patterns, which reflect their orientation to military service.

Changing professional identities have been explored in surveys periodically administered to officers over the last ten years (n= 5,000).[17] During this time, approximately 40 to 50 percent of junior officers have reported consistently that they "normally think of themselves as specialists working for the Air Force rather than as professional military officers." Although this finding may not seem surprising, further analysis reveals several surprises. First, this identity ratio of 60 percent officers and 40 percent specialists among junior officers holds true even in a sample of Air Force Academy graduates, who presumably should be more institutional overall. Second, in several surveys, flying officers demonstrated the greatest tendency to adopt a specialist identity and to view themselves as professional pilots who just happen to be flying for the government. This finding contradicts the assumption that the most institutional elements are found close to the flight line. Third, when I asked support officers why they identified themselves as officers, many replied, "Because I do management." The military nature of this identity must be questioned because management is not an inherently military function. Thus, both groups have adopted civilian norms and structures to give meaning to their military service.[18]

It is important to measure identities because they organize individual attitudes and values and provide a normative basis for committed action. In large samples, these identities explain consistent differences in a cluster of significant attitudes. Those who have officer identities, for example, tend to report as follows:

• they view military experience as a way of life, not as a job;
• their air force careers provide better opportunities for interesting and challenging jobs than would civilian careers;
• getting comparable jobs (in terms of importance) would be very difficult if they left the air force today;
• the air force does not require them to participate in too many activities not related to their jobs;
• personal interests must take second place to operational requirements for military personnel;
• air force people are special;
• they live on base rather than in the civilian community;
• they plan to continue their military service for 20 years or beyond.

By contrast, individuals with specialist identities reported the opposite attitudes (disagreement with the above statements) and indicated less interest in air force careers.[19]

Further analysis of these identities, accomplished with interview data from a smaller sample (n = 83), suggests important relationships between identities and commitment behavior. That is, the differences in identity clearly reflect the institutional or occupational orientations of military members.[20]

Persons with officer identities tend to see their jobs as involving several diverse activities that, taken as a whole, serve an important function for society. They view their service over an extended time, placing more importance on career than on job satisfaction. Their social interaction with the military is widespread, involving persons of different specialties and ranks. In sum, their involvement with the military is value oriented, broad based, and long term. They see their military jobs as special, transcending self-interest, and they regard the organization as a closely knit community. The personal investment patterns of these officers follow this value-oriented perspective. Because they place importance on the difference between military and civilian society, they tend to invest heavily in the military subculture: they work long hours at uniquely military tasks; they live in government housing; and the spouses work for the military community. They also tend to shut out civilian opportunities: they do not invest in particular geographic areas; spouses do not pursue careers; advanced education is not pursued seriously. Officers with this identity still view the military as an institution and their job as a calling.

Specialists have a different perspective. They place primary importance on their specialized function, which could be accomplished just as well in civilian organizations, and on satisfaction associated with jobs in that specialty. Their social interaction, which is limited, is usually related to shared interest in the specialty and is just as likely to occur in civilian as in military circles. Because this group places importance on the similarity of military and civilian work within the specialty, they direct their efforts toward generating alternative opportunities outside the military. When the cost of staying in the military is too high, perceived opportunities outside the military determine career decisions. Essentially, their career decisions tend to follow the economic model of maximum payoff among alternatives.

Thus macro-level social changes have predictable consequences for organizational change on the meso level and affect identity and commitment patterns at the micro level. Occupational trends that reach the micro level may be irreversible because their root cause, macro-level social changes, cannot be affected by those inside the organization (the meso level) and because cohorts of individuals affected (on the micro level) by these changes increase their influence as they move up in the military system.

IMPLICATIONS

In regard to the theory of I/O trends, reconceptualizing the I/O thesis as the complex interaction of social organizational change on the macro, meso, and micro levels suggests a fundamental rethinking of the focus and method used to explore these trends. Because the I/O thesis is essentially a multilevel process theory, we must focus on the process rather than on the outcome, and inquiry at a single level is not sufficient to illuminate the process. Further, scientific inquiry at any level must be appropriate for that level.

The case of the air force highlights the interactive nature of this process and suggests that several key analytical variables are driving social change in either an institutional or an occupational direction. The most important macro-level variable is *special status*; differences between the military and the civilian organization must be recognized and legitimated by the unique and important function of the military in society. At the meso level, *functional integration* is crucial. Shared activity, function, or vision must serve to integrate and focus individual effort toward the unique purpose served by the institution. At the micro level, the critical variable is *orientation of the members*. Institutional members orient themselves to the corporate whole and to the mission they perform, whereas occupational members tend to identify with their civilian counterparts.

The I/O thesis is not a prediction of occupationalism; rather, it offers a way of understanding ongoing social organizational change. The pressure for occupationalism is great because it stems from macro social change, which is articulated through the meso organizational level to the micro level. Left to operate unchecked, this trend is probably strengthened by a similar interaction from the micro level upward. As cohorts of members with an occupational orientation move through the military systems and gain influence, they will create meso-level organizational changes that fit their occupational orientation to military service and reduce further the distinctiveness of the military as a unique institution in society. Thus, social organizational changes at all levels interact in both directions.

Viewing the I/O thesis as a complex of multilevel changes in social organization not only suggests that we reconsider the underlying theory and method of investigation, but also gives us a new perspective on current organizational issues and informs future policy formulation. Essentially, an understanding of institutional change offers a perspective for effective policy formulation and a way to deal with strategic issues.

The professional identities and commitment patterns emphasized in this perspective, for example, may explain attrition rates. Because specialists respond to economic incentives and alternative opportunities, their attrition will be low (they will appear to be committed when military pay is comparable to civilian pay or when few alternatives are available in the civilian sector, (e.g., during a recession). When pay and bonuses are not competitive or when the economy improves, the attrition of specialists will increase. Officers, on the other hand, respond to normative differences. Above a minimum standard of living, they accept and even expect personal cost and hardship, as long as they perceive themselves as working for the collective good of society. When that goal is lost or when they are thrust into a situation in which the normative goal is self-interest or individual economic reward, their commitment decreases or they adopt the economic orientation characteristic of the specialist. Thus, an unchecked trend toward occupationalism on the macro and organizational level has an unexpected cost. It forces those with an officer orientation to reconsider and adopt a specialist orientation or to leave the service because the normative difference no longer exists. At the same time, those with a specialist orientation will require increasing economic incentives or will leave the service when alternatives in the civilian sector improve. In both cases, the military will become more occupational, attrition will be high, and the cost of military service will increase in the cash-work nexus of the marketplace.

This perspective also suggests some overall strategies for dealing with I/O trends. If these trends toward occupational organization are viewed as dysfunctional, at least two approaches are possible.

One option is to attempt to correct the situation by formulating policies and implementing changes that promote institutional structures, norms, and orientations. The air force has instituted at least two such programs. The first is Project Warrior, which attempts to highlight the combat function of the air force and to remind all members of the combat duties they might be asked to perform. Typically, Project Warrior programs involve military heritage weeks, combat uniform days, movies and posters, and special exercises highlighting the individual's part in the combat function of the air force.

Another program supporting this strategy to revive institutional values is the renewed emphasis on leadership rather than management relationships between superiors and subordinates. Management relationships are assumed to be less personal, less caring, and more characteristic of contractual relationships than of value-oriented relationships. Leadership is assumed to be more personal, more oriented toward shared goals and values, and more characteristic of organizations demanding loyalty and self-sacrifice for the greater good. In this regard, leadership is receiving more emphasis in formal training programs and is the subject of special publications available to all air force personnel. Unfortunately, both this effort and Project Warrior are sporadic and represent norms and modes of operation that conflict with the day-to-day operation of an occupational air force.

Although these are noble efforts to curb the occupational trend, they have a low probability for success. Part of the pressure for occupational organization and modes of operation comes from macro social changes over which the air force has little control. Assuming that any military organization can change the situation by itself may be unreasonable, especially given the recent loss of professional autonomy. Only when society views military expertise as unique and important will the military have the ability to organize and operate institutionally, that is, to be divergent from society in regard to structure and norms.

Another strategy that may be easier to implement, but that may create another set of problems, is to adopt a plural organization—part institutional and part occupational. In this schema, a small core of institutionally oriented military professionals would serve as the "managers of the instruments of violence." They would be selected (at about the 10-year point) from persons serving in all major specialties who have proven themselves to be outstanding performers and committed to a military career. Selection to this status would require them to move frequently and to accept a wide range of assignments to prepare them for management of the entire force; these assignments would involve operational, support, and reserve liaison duties. As part of this selection, their spouses would be required to relocate, forgoing the possibility of careers, and would be asked to support the social needs of the organization. In return for their support, perhaps, spouses could receive G.I. Bill benefits or their own old-age pensions.

Those not selected would be asked to serve as specialists. They would never be managers of the overall force (commanders authorized to commit troops to die), although they might serve as key supervisors in their specialty. They probably should not rise in military rank but could receive longevity pay increases and/or promotions to positions of greater responsibility. In addition, they could be allowed the flexibility to leave active duty and work part-time with the guard or the reserves and to return to active duty when air force staffing needs require and when their personal situations allow. Their obligation to the military could be specified by contract and would be more limited than that of the professionals, with at least a four-year period to amortize training or administrative costs.

This option is not a radical idea, but merely a formalization of actions the air force has already taken to ensure that critical positions are filled. Members on active duty, for example, have received special consideration and often have been released early if they join the National Guard or the reserve. Those with critical skills have been invited to return to active duty. Institutionalizing these programs would have several advantages, such as further integrating active and reserve forces, recognizing different orientations to military service, clarifying career contingencies, taking advantage of training acquired in the civilian sector that might be transferable, and providing a broad base of experience in the reserve, which could be used on short notice in time of war.

CONCLUSION

The I/O thesis articulates the long-term trend toward civilianization of the military that has occurred since World War II. This thesis can be conceptualized and measured as changing social organization on the macro, meso, and micro levels. On the macro level, society has redefined the meaning of military service, reduced military professional autonomy, and forced the convergence of military and civilian structure and function. On the meso level, these changes are reflected by a redefinition of the core function of the air force away from unique combat and flying activities and toward more general management functions. These changes have predictable consequences for micro-level changes—specifically, the civilianization of the professional identities and commitments of military members.

These trends toward occupationalism are complex but not inevitable. They can be reversed. If we understand how critical social organizational factors operate at each level and how changes at various levels interact, we note several possibilities for proactive intervention and effective strategies for institution building.

Institution building is fundamentally the task of organizational leaders at all levels. At the top of the organization, senior leadership (civilian and military) must mediate the changes at both macro and micro levels. They must take the responsibility for articulating what is distinctive about the military and for ensuring that the distinction is understood and legitimated outside and inside the organization. They must also shape values within the organization. Collectively, they must create a vision—a sense of organizational purpose—that is shared by all and that integrates all segments of the organization toward a common goal. Lower-level officers are also responsible for institution building. In their everyday interaction with the troops, *they are the institution*. The ultimate concern of every officer should be binding subordinates to the organization and to the mission. They must exemplify the values of mission over self and of devotion to the corporate body, even at the risk of their careers. Actions say more than words, and the troops know what is real and what is lip service.

The U.S. Air Force officer corps is on the leading edge of I/O trends, and leaders at all levels face some difficult challenges. Professional identities and commitments of air force officers are highly susceptible to occupational influences; although most officers still view themselves as military professionals, the changing nature of military service generates strong pressures to change their orientation and become more occupational or to leave the service. Institution-building efforts can make a difference, but they will demand hard work and the sacrifice of individual interests for the greater good. Decisions must be made about the desired mix of institutionalism and occupationalism and about the action required to achieve this mix. The military cannot be left to drift at the whim of social change.

NOTES

1. C. Wright Mills, *The Power Elite* (New York: Oxford, 1956), p. 195, and Samuel P. Huntington, *The Soldier and The State* (New York: Vintage, 1957).
2. Morris Janowitz, *The Professional Soldier* (New York: Free Press, 1960).
3. See Charles C. Moskos, "The Emergent Military," *Pacific Sociological Review* 16, no. 2 (1973): 255–288. See also Z. B. Bradford and F. J. Brown, *The United States Army in Transition* (Beverly Hills Calif: Sage, 1973); W. L. Hauser, *America's Army in Crisis* (Baltimore: Johns Hopkins University Press, 1973); and D. R. Segal, J. Blair, F. Newport, and S. Stephens, "Convergence, Isomorphism, and Interdependence at the Civilian Interface," *Journal of Political and Military Sociology* 2 (1974): 157–172.
4. See, for example, Jerald G. Bachman and John D. Blair, *The 'Military Mind' and the All-Volunteer Force* (Ann Arbor: University of Michigan Survey Research Center, 1975).
5. Charles C. Moskos, "From Institution to Occupation: Trends in the Military organization," *Armed Forces & Society* 4, no.1 (1977): 41–49.
6. Excessive loss rates of highly trained air force officers have been an important issue in recent years.

From 1978 to 1980, for example, pilots with six to 11 years of experience left the service at rates of up to 80 percent in some weapon systems. For a discussion of "doing more with less" see K. C. Stoehrmann, "The Do-more-with-less Syndrome," *Air University Review* 32 (November–December 1980): 103–108.

7. Morris Janowitz, "From Institution to Occupation: The Need for Conceptual Continuity," *Armed Forces & Society* 4, no. 1 (1977): 51–54.

8. Charles C. Moskos, "From institution to occupation: trends in the military organization," *Armed Forces and Society*, 4 no.1 (1977): 41–49.

9. For a more detailed explanation of these findings, see Frank R. Wood, "U.S. Air Force Junior Officers: Changing Professional Identity and Commitment," Ph.D. dissertation, Northwestern University, published as Defense Technical Information Center technical report # AD A119101, 1982, chapter 4.

10. See Huntington, *The Soldier and the State*, p. 28; and Bruce Russett and Miroslav Nincic, "American Opinion on the Use of Military Force Abroad," p. 147, and B. Guy Peters and James Clotfelter, "The Military Profession and Its Task Environment: A Panel Study of Attitudes," p. 60—both in Franklin D. Margiotta, ed., *The Changing World of the American Military* (Boulder, Colo.: Westview, 1978).

11. See Moskos, "From Institution to Occupation," 47–48; "Civilianize 50,000 Jobs Now—GAO" *Air Force Times* 23 October 1978, 10; and "Many Not in Service Are Eligible for Vet Benefits" *Air Force Times*, 12 October 1981, 6.

12. See G.T. Allison, "Military Capabilities and American Foreign Policy," *The Annals of the American Academy of Political and Social Science* 406 (1973): 17–37. For a classic example of how Robert S. McNamara rejected military expertise in favor of cost analysis, see D. O. Smith, "We'll Lose the Next War, Too, Unless," *The Retired Officer* 36, no.10 (1980): 18.

13. Past issues of the *Air Force Times* consistently report the frequency and the extent to which these agencies are active in day-to-day decisions.

14. A 1979 pilot study included unstructured interviews with 83 junior officers. A 1980 career commitment study surveyed 2,754 U.S. Air Force Academy graduates. A 1982 study surveyed the job attitudes of 3,162 U.S. Air Force officers, and a 1984 study surveyed the job attitudes and career decisions of 5,530 officers.

15. See Wood, "U.S. Air Force Junior Officers: Changing Professional Identity and Commitment," chapter 3, and "USAF Junior Officers: Prestige Leveling and Civilianization."

16. Franklin D. Margiotta, "A Military Elite in Transition," *Armed Forces & Society* 2 (1976): 155–184.

17. The sample sizes and purpose of these studies are outlined in note 14.

18. See Wood, "U.S. Air Force Junior Officers: Changing Professional Identity and Commitment", chapter 3 and "USAF Junior Officers: Prestige Leveling and Civilianization."

19. Data from 1984 survey of U.S. Air Force officers.

20. A more detailed analysis is reported in Wood, "U.S. Air Force Junior Officers: Changing Professional Identity and Commitment," chapter 5.

IV

The Institutional Organization Model and the Military

CHARLES A. COTTON

These are challenging and troubling times for defense planners faced with what Shakespeare called the "cankers of a calm world and long peace." As the détente era unfolds amid growing deficits and national dissent over the size and scale of military systems, debate grows over the shape of their evolution. The systematic evaluation of the appropriateness of the current—and evolving—defense arrangements is at best an art rather than a science, whether we are examining alternative modes of recruitment or unit operational effectiveness and individual combat commitment. Simply stated, protracted peace generates chronic debate over military goals, formats, structures, and cultural styles.

Much of this debate is really philosophy disguised as social and policy science argument. Although it is chronic, it is necessary and functional. In the words of John Downey, the British writer on military management, "the military is perhaps second only to religion in the extent to which, in peace, it must question its tenets and beliefs: in war it is second to none in its need for strong easily interpreted doctrine."[1] In this context, Charles Moskos's work on developmental transformation in military systems is significant in providing us with guidelines for institution building in the military.

This essay is concerned with the developmental institution-occupation (I/O) model of military change put forward by Moskos and its relationship to strategic issues of institution building in modern military systems.[2] Assuming that the readers are not concerned simply with an intellectual understanding of the ways in which military systems are developing, but also with insights into how to develop them, I argue that the I/O model is of considerable usefulness in any systematic study of institution building in the military. At a time when the great majority is concerned with the small pieces of the puzzle, Moskos challenges us to consider the whole and to pose deeper questions regarding the essential character of the military and its developmental direction. His model causes us to stand back from what Lewis Mumford termed the "aimless dynamism" of modern bureaucracies and to examine where we have been and where we are going.

The I/O model is more than a model, or "theory," put forward by social science to be "refined," "tested," and subjected to endless logical, conceptual, and methodological analysis and criticism. In one sense, it is definitely that, and examining it in that light is appropriate, especially given its policy implications. Yet the I/O model is not only descriptive and morphological; it is clearly normative in that it posits a desirable institutional state fading into the distance as time goes on. The I/O formulation is concerned manifestly with what *is* happening at both the macro and micro levels of the military, but it has a latent concern with what *ought* to be happening and how we can alter the lines of development. The normative dimension accounts for the widespread interest in this model and in its validity among senior policy makers. It captures the gut-level perceptions of many participants and observers regarding the essence of change in military systems.[3]

As a descriptive model of transformational change in military systems, the I/O formulation presents a number of problems that deserve serious intellectual discussion. This writer, among others, has raised several issues regarding the conceptual clarity and testability of the I/O formulation.[4] The ambiguity of terminology and causal postulations, the challenges of measurement at different levels of analysis, and the confusion of causes and organizational outcomes show ample room for "refinement." Much, if not all, of the empirical work using the I/O model has been concerned with attitudinal variance among military members—the micro subjective dimension of the model—even though it appears to be intended as a model of *system* change.[5]

I am concerned here with the model's essential relevance to the issue of institution building in the military. This task was chosen deliberately and it is not without its risks, but it is important if we are to attempt to set the I/O formulation in its broadest context. Because the attempt requires "divergent thinking" at certain points, I ask the reader's indulgence. We will turn briefly to one such divergence before considering the essence of the Moskos formulation.

AN INTERESTING PARALLEL

This seems an opportune moment to raise what I consider a highly relevant civilian parallel to the I/O approach: the *In Search of Excellence* (ISOE) phenomenon. Because I have used this book as a text in recent months, and because I have also been involved with the Canadian federal government in adapting the "attributes of excellence" to the public sector, I now regard the I/O formulation in a new light. At the risk of stretching a point, we can see the entire I/O literature as a search for the attributes of military institutional excellence. Just like successful and adaptive corporations with definable attributes, one may think of successful and adaptive military systems, although in peacetime the bottom line for the military is extremely blurred.

Of course, one could write a separate essay on this broad point. The I/O and the ISOE frameworks share a number of similar concerns. Both are prescriptive as well as descriptive; both are concerned with values and organizational cultures; both examine organization–environment relations; both focus on the corporate leaders' responsibility to create mission-oriented value systems for members; and both, interestingly, constitute attacks on the narrow, quantitative, management-techniques approach to organizational action and development. In addition, both strike a responsive chord in a wide readership in their respective segments of American society. Later in this paper I will argue that, in both frameworks, values and subjective perceptions (meanings) are the essential concern for organizational leaders and the essence of effectiveness.[6]

THE ESSENCE OF THE I/O MODEL

The I/O model, which deals with various dimensions of military service and organization, can be looked at in several ways. On the one hand, Moskos sets out to delineate the way attitudes toward military service, both inside and outside the military, are changing; on the other hand, he is concerned with the way organizational parameters, policies, structural arrangements, rewards, and culture are developing in tandem with those attitudes. Despite some ambiguity as to cause and effect, I believe that perceptual (attitudinal) changes predominate. This view is supported by the use of the term *organizational outcomes* to refer to the shift in attitudes toward an occupational model of military service and away from an institutional model, the two polar types in the I/O formulation. In other words, structure follows culture and perceptions.

Moskos argues that at least in the United States, a change is taking place in the way military service and participation are defined and conceptualized by key segments of society—those in uniform, civilian defense officials, and the civilian population. The traditional perception of military service as a calling, or vocation, to the nation by its citizens, legitimated by broadly based national values, is giving way to a subjective definition of military service as an occupation in the labor market, involving the performance of work for civilian types of rewards under specified contractual conditions. The traditional social perceptions of the military underlie the institutional model; the subjective definition underlies the occupational model.

Thus, unlimited liability in the service of the state is being replaced by limited liability in a unique federal department. Policies and organizational arrangements reflect increasingly this emerging social definition of military service. The organizational correlates erode traditional conceptions and make it difficult to sustain or return to the institutional model. In fact, these perceptions may become so prevalent and so much taken for granted that the developmental problem may go unseen. (In such a scenario, the transformation is not a hypothesis but a fact.) The organizational outcomes accumulate through time; they are never consciously orchestrated in terms of an organizational vision, but they add up just the same.

At the center of this transformation lie perceptions and values that shape choices in managing the culture of the military and its participants. For Moskos, these choices emerge in the context of policy formation within the military sphere, a process that is influenced by the interactive and interdependent relationship between the military and its host society. Society and the military may be analytically distinct, but the military is not a closed system, nor is it totally dependent on pressures from society. People make choices, however, and thus some degree of freedom exists in policy formation. The future that is emerging from organizational drift need not occur.

The I/O model focuses our attention on the ways in which military service and military organization are defined socially inside and outside the defense community. Social definitions shape policy choices that result in movement (i.e., the emergent military construct) toward or away from either of the two polar types. From my viewpoint, this approach considers the transformation of the entire military as an open system in the societal environment. This point deserves emphasis in view of the debate over the macro and micro aspects of the model. Although subjective perceptions, values, and attitudes (the micro elements) are explanatory variables, Moskos does not focus on the issue of individual values but on group and system perceptions inside and outside the military. This writer has measured individual values and role orientations but did so primarily to document the character of the military as a social system.[7]

As I pointed out above, the cultural focus of the I/O model extends beyond the boundaries of the military, although most of the research has focused on the perceptions and role orientations of military participants, particularly at the lower levels. Although this group is of policy interest, it is not the sole constituency implicated in the I/O model, which deals with the entire military in its national social context. Other key constituencies are the uniformed and civilian leadership, which debates and sets policy, the general public (a category that probably needs further analytic differentiation), and persons of recruitable age. We need to document the social perceptions of the leaders (generals, defense bureaucrats, defense intellectuals, and politicians) who choose an occupational policy over an institutional policy. Further, we need to triangulate these perceptions (policy orientations) with those of other key constituencies and to examine the contradictions involved.

My own research disclosed a clear example of the cultural paradox that emerges in the process of policy formation. The senior operational officers are almost uniform in their institutional orientation. Interviewed in private, they are extremely critical of policy trends toward the occupational end of the I/O continuum, but in their public bureaucratic roles, they are actors in that process.[8] Committed privately to the institutional model but publicly to occupational policies, they take out their frustrations in the officers' mess. I would argue that if they are serious in their commitment to the institutional military, they must act to put it in place and engage in military institution building according to their vision of what the military should be. Whatever the operational or support environment on land, air, or sea, senior military and civilian officials must take an active part in designing and implementing the organizational model and the underlying values they believe appropriate. I will discuss this action implication of the I/O model in the last part of this chapter.

At this point, summarizing my observations on the essence of the I/O model seems appropriate.

- The I/O model conceptualizes the military and society as interdependent and interactive. The military is clearly not a closed system.
- The social character of the military can be conceptualized as varying between two conceptual models—the institutional and the occupational.
- The social character of the military is composed of the values and attitudes of its members and of the particular policies and structural arrangements organizing those members.
- Values and social definitions of military service either as an institution or as an occupation influence the choice of policies and structural arrangements. The former shape the latter, although a recursive effect may be involved over time.
- The social definitions of military service among constituencies inside and outside the military system are significant in understanding processes of transformation in military systems.
- The military is changing through time away from the institutional model toward the occupational model; at the root of this change are changing social definitions about the *meaning* of military service in the national context.
- Those responsible for the long-range evolution and design of the military should grasp these transformational dynamics, consider their implications, and take action to alter the emergent character of the military. In this sense, the model has an implicit action imperative regarding policy formation and the shaping of values regarding military service. Military and civilian leaders in the defense community must learn to manage military culture.

The I/O model alerts us to the transformations occurring in the cultural definition of military service and to the outcomes of that transformation. The model was developed to describe and capture the broad sweep of change in the military sphere of the United States, but it is also applicable in other westernized industrial societies. Time and again the research of the past two decades has documented the "divided military" phenomenon.[9] At the center of each national case study lie changing and contrasting social definitions of military service, along the lines developed in the I/O model.

The key concerns of the I/O model are the social legitimacy of the military and military service within society and the cohesion and operational commitment within the military community. Values differ in the institutional and the occupational cases. The institutional model is seen—implicitly at least—to be high in social legitimacy, high in cohesion, and high in operational commitment. The occupational model, on the other hand, postulates low values in each case. Correspondingly, the effectiveness of the military—an ambiguous construct at best—is presumed to be high in the institutional case and low in the occupational case; the culture of the assembly line is not equivalent to the culture of the firing line. This proposition (it is a proposition rather than a fact) is central to the I/O model, and the implication is clear: we must build a culture of the firing line and make policy with this culture always in mind, working to strengthen the social legitimacy, cohesion, and operational commitment of the military and its membership.

This line of thinking about the institutional development of the military is not exclusive to the I/O model. James Fallows made a similar observation at the conclusion of his review of the American defense system:

> The most important task in defence is the one most likely to be overlooked since it lies in the realm of values and character rather than in quantities which can be represented on charts. Before anything else, we must recognize that a functioning military requires bonds of trust, sacrifice and respect within its ranks, and similar bonds of support and respect between an army and the nation it represents.[10]

I suggest that the institutional type in the I/O model parallels Fallows's vision of a "functioning military," while the occupational type, the product of "quantities on charts," is presumed to be less functional. Both internal and external bonds must be considered, however.

This dualistic focus is not always evident in other critiques of change in the military. A case in point is Savage and Gabriel's critique of managerialism and careerism (both presumed to be occupational correlates) in the American Military.[11] Their criticism is explicit rather than implicit, but their concerns for reform and development are centered *only within* the military. According to Savage and Gabriel, the military must be isolated from civilian values and society and must structure its culture and organizational format around the operational imperative of cohesion. This seems to be a narrow, internal focus, with no connection between military service and citizenship.[12]

The I/O model of long-run transformation in military culture provides a frame of reference for strategic policy development. In its broad outlines, it seems valid to this writer as a description of the shifts that are occurring and as a useful conceptual map of the social topography of the military. This model serves both as a source of hypotheses for empirical research on the changing morphology of military values and its organizational consequences and as a framework for policy action for civilian and military leaders concerned with shaping the development of the military in society. That process, which can be termed *institution building*, is the focus of the later part of this essay.

INSTITUTION BUILDING IN THE MILITARY: THE MANAGEMENT OF MILITARY CULTURE

A key issue in evaluating modern military systems as social institutions is the degree to which their cultural development is out of control. We must ask questions about the blueprint for organizational adaptation and the degree to which that blueprint of a desirable future provides for a measure of control over what Moskos would term the *emergent military*. In short, where is the military going? Who is shaping its development and according to what model?

The tentative answer to these questions, at least with regard to all-volunteer forces (AVFs), is not necessarily reassuring. The military seems to be drifting out of control toward the paradoxical shoals of isolation from society and unrestrained civilianization. A blueprint for institution building seems to be lacking, despite the enormous efforts in "policy development" in the personnel and organizational fields. Studies proliferate and policy choices are made, but little evidence exists that these articulate a vision of the type of military institution desired in the future. Changes accumulate over time into the broad shifts delineated in the I/O formulation to create internal organizational strains and a growing perception among leaders that something is "wrong" with military culture and values. The modal approach in the evaluation of management of military culture becomes reactive rather than proactive.[13] At times leaders appear to be on a runaway stagecoach, without a grip on the reins or a map of where they are going in relation to society.

Leaders consciously concerned with institution building in the military are by definition proactive in their approach to the management of military culture; they adopt strategies for institutional development consistent with a vision of a desirable future. The term *culture* as used here is taken from a recent paper by Edgar Schein, in which Schein defines culture as the "pattern of basic assumptions which a given group has invented, discovered or developed in learning to cope with its problems of external adaptation and internal integration, which have worked well enough to be considered valid, and therefore to be taught to new members as the correct way to perceive, think, and feel in relation to those problems."[14] This definition relates directly to issues of organizational values and perceptions of membership obligations, patterns of socialization, and social definitions of the right policies needed to make organizations effective.

Schein argues further that organizations must develop a consensus among members regarding key problems of external adaptation and internal integration. Externally, consensus must be developed on the mission of the organization, the goals that reflect this mission, the means of attaining these goals, and the criteria for evaluating effectiveness. Internally, leaders must work toward developing cohesion and value consensus. Where consensus is limited, organizational strains exist and questions arise about the legitimacy of the organization.

Thus military and civilian leaders of the military have a corporate responsibility to manage culture, and in this context of action the I/O formulation can provide a useful frame of reference.[15] Leaders (to use the terminology of *In Search of Excellence*) must develop a philosophy related to the core mission of the military and act as value-shaping leaders. They must seek to develop a hands-on, value-driven organizational style. Policies and strategies must be consistent with the vision articulated by leaders, and leaders must build commitment to that vision among members at all levels.[16] In I/O terms, the key question is whether that vision approximates the institutional or the occupational image of military culture.

Military culture must be managed through systematic institution building by military and civilian leaders who are conscious of their corporate responsibility, and this need takes on critical significance in periods of rapid change and growing ambiguity regarding the goals of the military. In periods of national threat and protracted operations, military service has broad legitimacy and public status, and the functionality of traditional military culture is clear. Decades of détente, however, coupled with rapid social and technological change, weaken both the external and the internal integration of the military institution in society, creating a special leadership challenge. The definition of military service as an occupation, with its various organizational outcomes, is clearly a product of extended détente.

Control over the social development and management of military systems is an issue in any discussion of strategies for institution building in the military. One must presuppose that the ability and willingness to control developments and shape culture exists; otherwise, such a discussion is a purely academic exercise without practical policy implications. Clearly, however, there are limits—which vary across societies—to the degree of control that can be exerted. In liberal democracies with voluntary military participation, the shaping of future developments must operate within traditions of democratic control. At the same time, the pursuit of strategies will be slowed by bureaucratic inertia and the need, in some instances, for consensual action by political, military, and civilian leaders. Some strategies can be pursued within the military by service leaders, but others need a broad sociopolitical consensus.

A FRAMEWORK FOR INSTITUTION BUILDING IN THE MILITARY

By itself, the I/O formulation does not provide an explicit and systematic conceptual approach for institution building in the military. Because it is intended primarily as a coherent description of transformational change, it serves, insofar as it is valid, as a picture of where the military *is* going, not where it should be going. (Again, however, we must recognize that the picture contains the implicit criticism of the direction of change.) This model does not systematically develop an image or a vision of a military institution adapted to the social and operational vagaries of industrial democracies in the late twentieth century.[17] It serves to stimulate thinking but does not structure action in a coherent fashion. To develop structure we must go a step further, but as we shall see, the hallmark concerns of the I/O model—social legitimacy outside the military, and cohesion and commitment within—provide the basic conceptual anchors.

A sociological concern with institutional development in the military requires the examination of the interdependent issues of the link between the armed forces and society, and the internal culture and effectiveness of the miltary as a unique social system with combat goals. It is necessary to develop a blueprint for shaping both convergent and divergent trends to create an emergent military that is operationally effective as a deterrent force and firmly embedded in a society in which the public perceives military service as a valued role in adult political life. In the long run, institution builders in the defense community must seek both high internal and high external integration.

These objectives are not mutually exclusive and contradictory, as many analysts would have us believe. I do not agree with those who advocate a focus solely on internal cohesion and traditional martial values, in which the military is treated in isolation from society. Many of the staffing and attitudinal problems of volunteer militaries are linked to qualitative trends in recruitment, as the middle class shuns service and marginal citizens become problematic service members. At the same time, however, there are enormous risks in concentrating solely on building links with society and adapting to social trends without considering the related issues of operational effectiveness and individual commitment. Somehow we must pursue the middle road, reconciling apparent opposites with a higher order of understanding.[18]

Leaders of the defense community must break away from the trap of zero-sum thinking about directions in development. An attempt to increase internal cohesion does not always imply a parallel decrease in the link with society. A cohesive and committed military does not have to be kept away from the "contamination" of civilian values and images; dedicated members who have internalized the military ethic need not pursue their careers in splendid isolation on posts and bases, supported by their spouses. Similarly, we need not assume that attempts to strengthen links between the military and society lead always and irrevocably to weakened commitment and operational effectiveness within the military.

INTERNAL AND EXTERNAL INTEGRATION OF THE MILITARY

In considering the issue of institution building in military systems, focusing on the idea of integration seems appropriate. Despite its definitional ambiguity, integration implies a drawing together of disjunctive elements in society; as such, it is related to the analysis of the relations between the military and society and between components within the military. In this regard, the distinction between external and internal integration is particularly useful. External integration refers to links between the military and society, and implicitly to social legitimacy, and internal integration refers to links within the military, and thus implicitly to the issue of internal cohesion.

Employing the traditional division into high and low levels along each dimension, we obtain a two-by-two matrix that is useful in organizing our thinking about the character of a specific system and its future development. Figure 4-1 shows these possibilities, which are rudimentary ideal types: (a) a military low on both internal and external integration; (b) a military low on internal integration and high on external integration; (c) a military high on internal integration and low on external integration; and (d) a military high on both dimensions. A specific military can be located conceptually in this matrix; depending on where analysts feel it "should" be, strategies can be developed. This matrix also can be linked conceptually with the I/O continuum, as I will discuss briefly below.

Although this conceptual matrix can be used to interpret any national military system, the characteristics vary considerably, especially between conscript-citizen systems and AVFs. For the present discussion, I will concentrate on AVFs in Canada and the United States, because the pressures—and trends—toward the occupational model are most pronounced in these systems.

Where one places current AVFs depends on one's frame of reference and, to a certain extent, on the pressure to defend the status quo. As Moskos and Lissak have shown, identifying indicators of convergence and divergence in current trends is possible.[19] A review of indicators, however, might disclose one classificatory alternative rather than another and that in the long run AVF development must be toward the high-external, high-internal model has already been argued.

In my search for the middle ground between those who see the AVF becoming more isolated from society and those who see the traditional institutional character being eroded by civilianization, I am pushed toward the conclusion that the current AVF approximates the low-internal, low-external type. This is a specialized federal bureaucracy with weak and ambiguous ties to society, while at the same time it exhibits internal dissent and strains. Its links to national values and social fabric and its internal cohesion have both suffered in recent decades.

EXTERNAL INTEGRATION
Social and Normative Links Between the
Military and Society

		Low	High
INTERNAL INTEGRATION Social and Normative Links Within Military	Low	The current military? (1)	The "convergent" image of a civilianized and ineffective military (2)
	High	The "divergent" image of an isolated, cohesive military (3)	The future military? (4)

Conceptual Links to I/O Model
(1) True Emergent Occupational Model.
(2) Pseudo-Occupational Model.
(3) Pseudo-Institutional Model.
(4) True Institutional Model.

FIGURE 4-1. Developmental Images of the Military

In terms of the I/O formulation, the occupational image of the military is closest to this low-external, low-internal model, and the institutional image is closest to the high-external, high-internal models. Moskos argues that the occupational military has no special significance within society, even though some of its policies may have civilian parallels, nor does the internal culture of the occupational military have a unique and integrated character. This pattern differentiates it from the low-internal, high-external convergent image, which I regard as the pseudo-occupational type. In the same way, the high-internal, low-external divergent image is the pseudo-institutional type. Superficial developmental analysis focuses on these two categories and on movement along this axis, but this focus represents a misreading of the I/O formulation. Military-civilian links in both values and structures are essential ingredients in Moskos's institutional model. Military service in the institutional model, for example, is legitimated by national values and national goals, not by individual occupational aspirations.

If we consider several indicators of internal and external AVF integration, the case for the low-internal, low-external categorization is sharpened. (One cannot "prove" the case; to do so would require major research, and the results would still remain open to controversy. In a similar way, one cannot "prove" the I/O thesis.)

Indicators of low external integration include the following:

• middle-class antipathy toward military service

- narrow marginal recruitment base
- generally low participation rates
- sociopolitical alienation of junior troops
- weak integration between military service and civilian educational systems
- withering away of reserve systems

Indicators of low internal cohesions include the following:

- internal attitudinal conflicts
- ambiguity over the status of warrior
- strain between operational and support segments
- estrangement of experienced officers from policies
- variation in pay by occupation
- personnel turbulence in units
- careerism and turnover

My reading of the literature suggests that these AVF integration indicators are similar for Canada and for the United States. The degree to which they are evident in other national contexts cannot be assessed here.[20]

The above list suggests initiatives for institution building with regard to the AVF format. In terms of the conceptual argument in this paper, room for improvement exists along both external and internal dimensions. The process of institution building requires that we be concerned with strategies aimed at both dimensions; the blueprint must avoid the trap of zero-sum thinking.

I do not intend to discuss a wide range of strategies focused on the indicators listed above; I offered several such strategies in an earlier article.[21] Here I will center attention on two broad concerns that reflect the dimensions of this conceptual matrix: external and internal integration. In doing so, we must recognize that we cannot turn the clock back to an imagined past by adopting a few discrete programs and policies consistent with the institutional model of a military displaying high external and internal integration. Instead, the central concerns are the broadening of the military participation base and the management of corporate values.

Military Participation

The immediate problems faced by AVFs focus on the need to broaden participation in military staffing programs (an external strategy) and to clarify and delineate the core values of the military institution, including the essential meaning of military service (an internal strategy). To pose these problems in these terms is to go beyond a simple concern with the numbers of "qualified" military personnel who fulfill their "work" obligations in the defense bureaucracy, and to address issues related to the military as a national institution. Again I/O formulation sensitizes us to the deeper implications of how issues are defined and to the images of military service that underlie those definitions.

On the dimension of external integration, the immediate adaptive problem for AVFs is management of the recruitment challenges caused by demographic shifts and by increasing competition between military service and educational opportunities. (This problem has a higher priority than staffing the reserves.) Expanding the recruitment base by working toward more national representation in backgrounds is necessary. At present, middle- and upper-middle-class youth do not participate proportionally, and the quality of recruits is problematic, creating high training costs and first-term attrition.

This challenge is reinforced by the definition of military service as a job and a career rather than a stage in transition to political adulthood in a democracy. The perception of military service in occupational rather than institutional terms limits the range of policy options; the educational system, itself facing declining numbers, beckons middle-class youth as a career stepping stone. The emphasis on wages as the incentive for attracting a career force tends to perpetuate the existing patterns of marginal recruitment.

Although a future objective might be the acceptance of national service (the ideal expression of external integration in Moskos's institutional model), short-run strategies must be found that increase the integration between military service and educational institutions and that make some military service meaningful to the middle class. This approach is currently being developed in Canada and has been advocated by Moskos in the United States.[22] Essentially, it involves a shift from a professionalized career force as the ideal concept to acceptance of a two-tier system as the foundation for national military personnel procurement. A social fact of life in liberal democracies seems to be that most youth will find a military career unattractive, regardless of the economic incentives. On the other hand, many young people might find a short military tour attractive, with appropriate incentives. If participation in this first tier of military service were linked to educational benefits, as Moskos advocates, middle-class involvement would be likely to rise.

One variant on this strategy has recently been implemented in Canada. In this program, military technicians are trained in civilian vocational schools in two- and three-year courses. Civilian institutions have shown interest in adapting programs to meet military requirements; the students are full members of the military throughout their courses and spend their summers training in military settings. At the end of their program, they enter the enlisted ranks at a higher level than a new recruit and fulfill an obligatory term of service.

This program is unique because the enlisted personnel are recruited through a strategy that links military service and civilian educational opportunities. The program creates links between the military and society in a number of complex ways; civilians and military members interact, and civilian education authorities become sensitive to aspects of military service. The training bill for the military is also reduced.

Whether recruits join the military for subsequent educational benefits or are attracted to civilian schools by the prospect of military benefits is not really at issue here: in both cases the result is increased integration between the military and society. I suspect that this form of institution building will increase toward the end of the century, in spite of resistance by advocates of the pseudo-institutional image who decry benefits offered to those who have not "started at the bottom." Again, the specter of zero-sum thinking looms large.

Such strategies confront directly the fact that middle-class life and military service are disjunctive in the AVF format. The great majority of middle-class youth enter adulthood without experience, and their perception of it is unfavorable in most cases.

Institutional Values

Although strategies for external integration can increase the numbers and quality of persons coming forward for military service, they do not address the meaning of military service and the need to develop a coherent value system for AVF participants. This is a matter of internal integration; the second challenge for military and civilian leaders is to articulate a definition of military service and a core value system that will reduce the ambiguity and the structured strain within the armed forces. To state this issue somewhat differently, a coherent philosophy of military service is needed, developed at the top and put into action in the daily life of the institution. Leaders must debate and clarify consensual values and core role definitions and take action as *value-shapers*, to use a term from *In Search of Excellence*. In short, they must articulate what I have termed the *military ethos*.[23] They can no longer take internal cultural integration for granted but must act explicitly to shape it. The I/O attitudinal research confirms this institution-building requirement.

At the very least, leaders must clarify the meaning of military service and set the scope and limits of role obligations for service members at any level of the bureaucratic hierarchy. They must create and sustain a corporate moral code for the armed forces and its segments. If that moral code, with its core definition of the meaning of military service and its obligations, is to be characterized by a sense of military service as an individualistic career in a government bureaucracy, with limited performance liability, then so be it.

The challenge is to articulate a military ethos that is close to the essence of the institutional image presented by Moskos. At the outset, this ethos would define military service as an unlimited liability in service of the national interest and delineate the desired culture of the military community. This is an enormously complex task, and one which officers are reluctant, by training and inclination, to confront. Yet, if they do not take steps to clarify and shape values, they will remain at the mercy of I/O dynamics. In a context of institutional drift, broad social forces will shape values by default.

The articulation of an institutional ethos for the military has obvious links to issues of professionalism, ethics, socialization, and the allocation of status and rewards. It can also provide a frame of reference for policy development, leading to certain organizational outcomes rather than others. The key point, however, is that the shaping of values and perceptions takes precedence over the development of policies and programs. This is the essence of the I/O argument: images of military life influence policy choices.

Attempting any specification of a military ethos for AVFs would be inappropriate here; one can only offer some general observations regarding a desirable emphasis. National contingencies and traditions vary considerably but whatever the specifics, the emphasis would be on a morality of military service that is

> nation-centered, not organization-centered;
> mission-centered, not career-centered;
> group-centered, not individual-centered;
> service-centered, not work-centered.

Traditionally, these values have been emphasized in military creeds and culture. Military leaders have always sought to bring them to life in the armed forces, and the best leaders are still seeking them today.

These values should be emphasized in institution building along the internal integration dimension. If we avoid the question of values and remain content to debate the relative merits of institutional and occupational policies, we run the risk of contributing to organizational drift and of allowing our nostalgic visions to draw us back to the good old days.

Policies and programs are always subject to revision; the organizational features adapted to one era seem ludicrous in the next. The challenge is to articulate and sustain the appropriate values and perceptions, and *then* to evaluate policy alternatives in terms of those values and perceptions. As Peters and Waterman put it in *In Search of Excellence:* "If you get the values right, then the other things fall into place." I believe that the fundamental message in the I/O formulation is that military and civilian leaders in the defense community must get the values right. If this essay helps clarify that collective obligation, it will have served its intended purpose.

NOTES

1. J. C. T. Downey, *Management in the Armed Forces: An Anatomy of the Military Profession* (London: McGraw-Hill, 1977), p. 62.
2. See Charles A. Cotton, "The Institution Building in the All-Volunteer Force," *Air University Review* 24, no. 6 (September–October 1983): 38–50.
3. I know of no data to support this hypothesis, but I am convinced of its validity. In Canada, at least, the perception is widespread among experienced officers that traditional conceptions of military

service are being eroded and civilianized. Senior departmental officials have publicly observed that the forces may be facing a "crisis of the military ethos." See the discussion by Peter Kasurak, "Civilianization and the Military Ethos: Civil-Military Relations in Canada," *Canadian Public Administration* 25 (Spring 1982): 108–129. The I/O model receives considerable interest among nonacademics because it speaks to issues of effectiveness, strategy, and organizational adaptation in terms that senior military officials readily comprehend. See L. L. Cummings, "The Importance of Processes and Contexts in Organizational Psychology," *Graduate School of Business, University of Wisconsin, Madison, Office of Naval Research Technical Report* 1-1-3 (September 1980).

4. Charles A. Cotton, "Institutional and Occupational Values in Canada's Army," *Armed Forces & Society* 8, no. 1 (Fall 1981): 99–110. Of particular relevance here is Morris Janowitz's "From Institutional to Occupational: The Need for Conceptual Clarity," *Armed Forces & Society* 4, no. 1 (Fall 1977): 51–54. For research that represents the most systematic attempt to examine the attitudinal dimensions, see Michael J. Stahl, T. Roger Manley, and Charles M. McNichols, "A Longitudinal Test of the Moskos Institution-Occupation Model," *Journal of Political and Military Sociology* 9 (Spring 1981): 43–47.

5. On this point, consider the studies cited in the preceding footnote. The specification and operationalization of systemic variables is a challenging task, and Moskos has offered little guidance. This writer suspects increasingly that he and others are measuring the wrong things and the wrong people. Instead of documenting the attitudes of lower-level participants in the military, we should study those who shape policy and the perceptions of military service and organization that underlie their actions. We need to study the actions of the elite, most probably through the case method. This study would be problematic: it is difficult to enlist very senior defense officials as subjects in a study, and creative in-house research is rare. Thus, we are likely to see a continuation of research into the value orientations of lower- and middle-level participants, rather than those who are nominally in control of institution building. The situation in Europe, however, is different. See Jurgen Kyhlmann, ed., *Military and Society: The European Experience* (Munich: Sozialwissenschaftliches Institut der Bundeswehr, 1984).

6. Thomas J. Peters and Robert H. Waterman, Jr., *In Search of Excellence* (New York: Harper and Row, 1982).

7. Cotton, "Institutional and Occupational Values," pp. 99–110.

8. This phenomenon, which represents the real paradox of institutional development in the military, continues to perplex this writer. Those nominally in control of shaping policies through action in their public roles are privately extremely critical of the direction of change. Senior officers, especially from operational classifications, typically blame civilian military planners and the "civilians-in-uniform" who treat the military as a job. This phenomenon, which I term the *beleaguered warrior syndrome* is discussed at length in Charles A. Cotton, *The Divided Army* (Unpublished Ph.D. dissertation, Carleton University, Ottawa, 1980).

9. See, for instance, Cotton, *The Divided Army*, chapter II, for a cross-national review of research on attitudinal bifurcation in military systems.

10. James Fallows, *National Defense* (New York: Random House, 1981), p. 171.

11. Paul Savage and Richard Gabriel, *Crisis in Command* (New York: Hill and Wang, 1977). Also see William Hauser, *America's Army in Crisis* (Baltimore: Johns Hopkins University Press, 1973) for an approach that argues for a segmented military, composed of a "support army" adapted to civilian society and a "fighting army" isolated from it. Hauser's approach has the advantage of being immensely logical and the disadvantage of being socially divisive. One suspects that militaries throughout history have included some elite fighting units differentiated from the masses.

12. The need for military institutions to reflect the core values of democracies has been a major theme in the work of Janowitz, Sarkesian, and others. For a statement by a distinguished American soldier, see Robert Gard. "The Military and American Society," *Foreign Affairs* (July 1971): 1–13. Also relevant here is the argument by Stephen Westbrook, "Sociopolitical Alienation and Military Efficiency," *Armed Forces & Society* (Winter 1980): 170–189.

13. The military and its leaders appear to be "buffeted" by the impact of change, and the social consequences of new technologies are rarely anticipated. Conventional wisdom leads soldiers to prepare for the previous war, and most of their efforts seem dedicated to preserving old institutional patterns rather than building new ones. The danger with the I/O formulation is its superficial appeal to traditionalists who seek an academic justification for the old ways of doing things.

14. Edgar H. Schein, *Organizational Culture and Leadership* (San Francisco, Calif.: Jossey-Brass, 1985).

15. Chester Barnard, in *The Functions of the Executive* (Cambridge, Mass.: Harvard University Press, 1968), argued that, in addition to technical competence, leaders had the obligation to manage corporate morality and the core values of the organization and to energize members to accept them. This obligation is implicit in all traditional models of military leadership: military leaders are "accountable" for the morale (morality) of those under their command.

16. See Peters and Waterman, *In Search of Excellence*, chapter 9.

17. The confusion between the descriptive and the prescriptive tendencies in the I/O model is a stumbling block to its policy relevance. The model is explicitly descriptive (even if dimensions of its ideal types are blurred at times), but implicitly prescriptive. This ambiguity allows the reader to impose his or her biases in choosing to applaud or decry the direction of change in the military. The evaluation of trends, however, requires a model or vision of a desirable state of affairs; otherwise we never know whether things are getting better or worse. Note, however, that Moskos considers occupational ascendancy in a negative light in a subsequent policy-oriented article: Charles Moskos, "How to Save the All-Volunteer Force," *The Public Interest* (Fall 1980): 79–89.

18. See Gard, "The Military and American Society," on the need to find a middle road between the two extremes of isolation and civilianized ineffectiveness.

19. Charles Moskos, "The Emergent Military: Civil, Traditional, or Plural," *Pacific Sociological Review* 16 (1971): 255–280. See also Moshe Lissak, "Some Reflections on Convergence and Structural Linkages," paper presented to the Ninth World Congress of Sociology, Uppsala, Sweden, 1978. Lissak's paper points up the danger of simplistic assumptions about social change and the AVF.

20. There are obvious difficulties in considering AVFs and citizen military systems together in relation to the I/O concept. The choice of an AVF format in itself indicates erosion of the institutional concept of military service. The moral nexus of national service is replaced by the cash nexus of the marketplace, but the preexisting values remain within the organization. One might say that in AVFs the gap between ideology and reality increases.

21. Cotton, "Institution Building in the All-Volunteer Force."

22. Moskos, "How to Save the All-Volunteer Force."

23. Charles A. Cotton, "A Canadian Military Ethos," *Canadian Defence Quarterly* 12, no. 3 (Winter 1982/83): 10–19.

V

The Social Psychology of Military Service and the Influence of Bureaucratic Rationalism

JOHN H. FARIS

This chapter addresses two distinguishable dimensions of the social psychological issues inherent in Charles Moskos's formulation of the institution-occupation thesis (I/O—the attitudes, perceptions, and corresponding behavior of military personnel; and the beliefs, theories, methodologies, and rhetoric of military analysts, researchers, and policy-makers.[1] Most analyses of the I/O question have concentrated on the former. I argue that the latter is primary and more important. In addition, this chapter attempts to develop the historical and sociocultural context of these issues and discusses the implications, in terms of effectiveness, of current directions of change in military organization that the I/O thesis seeks to identify.

THE ESSENCE OF THE I/O THESIS

Many words have been spilled by now over what Moskos really meant when he introduced the conceptual apparatus of "institution or occupation." The catalytic article was immediately critiqued by Moskos's friend and colleague Morris Janowitz on grounds of insufficient clarity, and subsequently numerous commentators have been unshrinking in telling Moskos himself what he really meant, or what he should have meant.[2] To some degree, Moskos succeeded in creating a monster that has exceeded his own ability to

57

control. No doubt this is because his original analysis struck a nerve that is connected to an important truth.

To try to cut through the clutter of competing and contradictory interpretations of doctrine to illuminate the heart of the matter, to single out that sensitive nerve and follow it toward the truth is therefore vital, though hazardous.

We can begin by agreeing that the Moskos formulation fundamentally intends to identify and describe an important process of change in the nature of military organization.[3] What is problematic are the questions of the driving mechanism and the consequences of change.

Moskos's original 1977 article suggests an answer to this first question: "A shift *in the rationale* of the military toward the occupational model implies organizational consequences in the structure, and perhaps, the function of the armed forces."[4] His discussion of this point makes it clear that, although antecedents can be identified, the watershed for this process of change was the approach taken by the Gates Commission in planning the transition from conscription to the all-volunteer force, an approach that relied on "monetary inducements guided by marketplace standards." Thus, we have a clear formulation in this excerpt of the hypothesis that the shift in the rationale of the military precedes, and causes, the shift in structure and, "perhaps," in function. This formulation of the central element of the I/O thesis is a valuable starting point from which to disentangle the competing issues that surround it.

The Dangers of Literalism and the Need for Two Levels of Analysis

The contrast between institution and occupation is, of course, what gives the concept its cogency and power. One can enumerate a series of related contrasts. Institution is to occupation, in the sense Moskos means these, as—to some degree—self-sacrifice is to self-interest, cohesion and morale are to job satisfaction and relative deprivation, patriotism is to labor market elasticities, and sociology is to economics. Whatever truth value there is in Moskos's thesis, though, surely it is not that the military is moving from being an institution to becoming an occupation, in a literal sense.

Treating change in the military in terms of a literal shift from an institution to an occupation presents at least two problems. First, institution and occupation, in conventional usage, are not mutually exclusive. Some occupations—the professions—are clearly institutions as well, based as they are on internalized norms and values. A wide variety of other occupations (perhaps all) show at least some institutional characteristics. Long-haul truckers, magicians, schoolteachers—all have an ethos of established conventions, customs, and values. So if occupation is to be contrasted with institution, it must be a special form of occupation devoid of institutional qualities.

Second, to suggest that the military could be possibly become an occupation (whether institutional or not) is entirely absurd, An institution can have—and the military does have—a multiplicity of functionally differentiated roles organized in a relationship of organic solidarity. An occupation is essentially homogeneous, so the military could become only a cluster of occupations—of dental assistants, mechanics, tank gunners, and so forth—and accounting for the mechanism by which these discrete occupations are to be coordinated would be necessary.

Thus, what Moskos has referred to as *occupation* actually must be analyzed on two separate levels, which should henceforth be kept analytically distinct. The first of these levels, the "lower" level, deals with the tendency of some military personnel to think of themselves as having just another job, that is, as not being connected to the special norms and values that are peculiar and central to the military institution. For lack of a better term, and for the sake of some continuity with Moskos, I call this *occupationalism* (to be translated as *just-another-job-ism*). The second, "higher" level that derives from Moskos's concept of occupation describes the new rationale of the military that has, in part, displaced the traditional rationale and is manifest in the language, analysis, and decision making of any military researchers, analysts, and policy makers. This new rationale I will call, after Max Weber, *bureaucratic rationalization* for reasons to be detailed later.

The core of my argument is that, for a variety of reasons but especially because of philosophical and cultural factors to be discussed here in depth, the rationale of the military organization has shifted in the direction of bureaucratic rationalization; that this has in turn, both directly and indirectly, affected both the attitudes and values of military personnel as individuals (including some increased occupationalism) and the morale and cohesion of military units; and that these effects then have potentially severe deleterious consequences for the military's ability to perform its function. This set of propositions requires extended development.

I/O Tensions in the Civilian World

It makes sense to begin by considering the broader context of contemporary American society. Issues that parallel the I/O tensions in the military are not absent from civilian life. Rather, they occupy a dramatically central position.

Let us consider which civilian occupation best fits Moskos's ideal type of an occupation (as contrasted with institution).

We have already noted that institution and occupation in conventional usage are not mutually exclusive: various civilian occupations such as physicians are also institutions. In fact, identifying an occupation that fits Moskos's ideal-type characterization becomes difficult. One might suggest the commission salesman—the door-to-door huckster—as the apotheosis of the occupational model. But listen to Zig Ziglar:

> I Am a Salesman
>
> I am proud to be a salesman because more than any other man I, and millions of others like me, built America. . . . As a salesman I've done more to make America what it is today than any other person you know . . . I have educated more people; created more jobs; taken more drudgery from the laborer's work; given more profits to businessmen; and have given more people a fuller and richer life than anyone in history . . .
>
> Without me the wheels of industry would come to a grinding halt. And with that, jobs, marriages, politics, and freedom of thought would be a thing of the past. I AM A SALESMAN and I'm both proud and grateful that as such I serve my family, my fellow man, and my country.[5]

Whether Ziglar actually subscribes to such sentiments is not important. What *is* significant is that such books as Ziglar's, filled with enunciations of ethical codes and value statements, sell in the hundreds of thousands to people in sales, who also buy cassettes and attend seminars that communicate the same message.

Other possible illustrations of the ideal-type occupation are blue-collar factory workers and fast-food restaurant workers. However, Donald Roy (among many others) has shown how even the most routinized manual operations can be permeated by issues of meaning and how the social relations in factories can develop into highly completed subcultures.[6] Even fast-food restaurants can develop a culture that maximizes voluntaristic employee involvement in a process of continuous improvement of group performance and product quality.[7]

This is not to say that occupationalism does not exist in the civilian world, but rather to suggest the possibility that wherever one finds the rhetoric of pure calculation and the cash nexus, one is also likely to find some form of institutional reformation in an attempt to resurrect or introduce an element of transcendent meaning into the context of instrumental activity. From this point of view, then, one reason for the popularity of exhortational tracts such as Ziglar's within the sales field is that the apparent valuelessness of their work creates a serious hunger for reassurance. What also should be noted, though, is that these tracts invariably make a strong connection between attachment to meaningful values and effectiveness on the job as measured by promotions, earnings, and other occupational criteria. The "I Am A Salesman" manifesto, for example, appears in a book titled *Zig Ziglar's Secrets of Closing the Sale* and another of his best sellers is called *See You at the Top*.

Another, somewhat more sophisticated variant of institutional reformation in popular business literature is the spate of books such as *The Art of Japanese Management*.[8] *Theory Z*,[9] and the phenomenally successful *In Search of Excellence*.[10] In the last, Peters and Waterman frame the conflict between bureaucratic rationalism and institutional values quite succinctly:

> The numerative, rationalist approach to management dominates the business schools. It teaches us that well-trained professional managers can manage anything. It seeks detached,

analytical justification for all decisions. It is right enough to be dangerously wrong, and it has arguably led us seriously astray.[11]

Again, their message—that the business school mentality is not only morally barren but also ineffective—falls on receptive ears. All sorts of corporate advertising trumpet a "commitment to excellence," and Tom Peters has become an evangelical industry in himself. Interestingly Peters's follow-up book (with Nancy Austin), *A Passion for Excellence*, echoes the disenchantment of some officers with the military's emphasis on management and urges a revitalization of leadership as a substitute:

> We believe the words "managing" and "management" should be discarded. "Management," with its attendant images . . . connotes controlling and arranging and demeaning and reducing. "Leadership" connotes unleashing energy, building, freeing and growing.[12]

Peters and his colleagues are not without their critics, of course, but their widespread influence, as well as popularity, suggests that they, like Moskos, have struck a sensitive nerve. Some of the most pressing issues in the world of business today can thus be seen as grappling with I/O tensions that, to a degree, parallel those within the military.

The Military as a Special Case

Attachment to institutional values needs continuing affirmation and nourishment, and organizational effectiveness is always problematic. This point is especially pronounced in the case of the military because of both the extreme requirements occasionally imposed on military personnel and the severe consequences of organizational *in*effectiveness. Also, organizational effectiveness of the military is only sporadically tested, and within the context of nuclear deterrence, even these "tests" tend to be less than conclusive. (Does our ability to force down an Egyptian jetliner show that our armed forces are as prepared as they need be?)

I/O tensions in the military are undoubtedly as old as cash remuneration for soldiers, or older. Achilles sulked in his tent because he felt Agamemnon received too large a share of the spoils, though Achilles had previously epitomized the warrior ethos.[13] Without question, I/O tensions in the modern military precede the all-volunteer force. In the nineteenth century, one could buy one's way out of the draft in the American Civil War and at the same time buy a commission into the British officer corps. The post–World War II U.S. military showed clear signs of I/O tensions, as indicated in Janowitz's 1958 interviews with 113 Army officers:

> Many a military officer sees his career as filling some special mission, rather than as just a job. Some effects of the transformed selection system can be seen in the career motives of potential members of the military elite. Those who see the military profession as a calling or

a unique profession are outnumbered by a greater concentration of individuals for whom the military is just another job.[14]

Occupational attitudes are especially disquieting to the military because the rhetoric of self-interest conflicts so dramatically with the demands for self-sacrifice and suppression of self-interest required for effectiveness in combat.

THE ALL-VOLUNTEER FORCE AND THE ACCELERATION OF OCCUPATIONALISM

As one might suspect from a review of the history of military organization, some degree of occupationalism is no doubt inevitable. In fact, as actual combat is experienced less and less frequently by a decreasing fraction of a military organization in a posture of deterrence, containing occupationalism within manageable limits becomes increasingly difficult.[15] A central proposition of this paper is that the bureaucratic rationalism (which subsumes economic theory and practice) on which the U.S. all-volunteer force was deliberately designed and implemented has dramatically fueled rather than dampened occupational attitudes within the military.

Consider a few examples of this bureaucratic rationalism in action:

[W]e feel there is some justification for a high-quality Army recruiting policy, based on rigorous cost analysis. It may actually be cheaper in some circumstances to replace low-quality accessions with high-quality recruits with superior attrition characteristics. We should emphasize that in all our analyses, we implicity assumed that high-quality soldiers are no more productive on the job than low-quality soldiers.

[A] wider variety of retention incentives is necessary to maintain the required mix of military careerists in the future. An advantage of our civilian job-growth-centered model is that it can help focus this retention research on specialties that face the strongest civilian sector competition in the 1990s.[16]

These examples require no comment. However, in "Economic Analysis of the All-Volunteer Force," McNown, Udis, and Ash present a counterattack to Moskos, defending the market model from claims that the all-volunteer force has failed to meet recruiting goals by asserting that, because pay levels have not been sufficiently competitive, the market model has not been given a chance to work.

If we take as a "best estimate" of the pay elasticity the .86 figure for the total Department of Defense accessions, to maintain the current level of accessions will require a 17.44% increase in military pay *relative to civilian rates of pay*.[17]

The willingness to calculate and report required pay increases to four digits is notable. What is more important is that though the authors quote Moskos ("The grievous flaw has been a redefinition of military service in terms of the economic marketplace and cash-work nexus"[18]), their defense does not address this point. Moskos's point, oversimplified, is *not* that you cannot buy enough recruits to meet manpower goals, but that if you do, you risk a potentially calamitous transformation of the meaning of military service.

That the authors cited above are economists is no accident. One reason why bureaucratic rationalism is preeminent in planning and implementing the all-volunteer force is that this is merely another manifestation of the pervasive ascendance of economics as the dominant social science and reigning model of the world. Sheldon Wolin is among the numerous observers who have commented on this ascendance:

> What can hardly be doubted is that economics now dominates public discourse. It is now common practice to rely upon economic categories to supply the alternatives in virtually every sphere of public activity, from health care, social welfare, and education to weapons systems, environmental protection, and scientific research; and to function as a sort of common currency into which all problems have first to be converted before they are ready for "decision-making."[19]

One of the most authoritative critiques of contemporary economics is that of Lester Thurow, himself a rigorously trained professional economist:

> The late 1960s and early 1970s became the age of economic imperialism.
> Economists filled journals with articles on the economics of crime, marriage, suicide, and other social phenomena once regarded as the exclusive preserves of other social sciences. Economists thought they had a better analytical apparatus for studying those phenomena than other social scientists, and what's more, their imperial claims were accepted by many other academics . . . To protect their flanks against the inroads of economists, other academics felt that they had to become more like economists.[20]

Thurow's point has special application to sociology, which has increasingly shifted toward a form of ersatz economics, with notable exceptions including most sociological work on the military.[21]

The Rise of Bureaucratic Rationalism and the Myth of Managerial Effectiveness

The rise of bureaucratic rationalism, both in the military and in civilian institutions, cannot be explained simply in terms of the ascendance of economics. A broader, more long-term context of sociocultural trends underlies this ascendance, and the particular strains of the contemporary military can be understood better if these trends are examined.

The current dominance of economics, including the rhetoric and policy direction that guide the management of the all-volunteer force, can be seen as one symptom of a more fundamental process: the long-term, attempt to replace religious faith and certitude with a secular rationality, a model of events that would provide order, understanding, control, and certainty.

Irving Kristol notes the displacement of traditional beliefs and values by "worldly rationalism":

That spirit of worldly rationalism so characteristic of a commercial society and its business civilization (and so well described by Max Weber and Joseph Schumpeter) had the effect of delegitimizing all merely traditional beliefs, tasks, and attitudes. The new, constructed by design or out of the passion of a moment, came to be seen as inherently superior to the old and established, this latter having emerged "blindly" out of the interaction of generations.[22]

As Edward Shils puts it, the orientation of the social sciences ever since the Enlightenment is a "skeptical attitude towards tradition and a conception of society which leaves little place for it."[23]

As a result, we find the same syndrome of a mechanistic, deterministic model of human behavior in fields outside economics—in sociology in the form of exchange theory and other fashions, in psychology in instinct theory and behaviorism, and, as the following analysis by John Keegan suggests, in military history:

What has been called the "rhetoric of history"—that inventory of assumptions, and usages through which the historian makes his professional approach to the past—is not only, as it pertains to the writing of battle history, much more strong and inflexible than the rhetoric of almost all other sorts of history, but is so strong, so inflexible and above all so time-hallowed that it exerts virtual powers of dictatorship over the military historian's mind.[24]

The key features of this rhetoric of history noted by Keegan include the assumption of "extreme uniformity of human behavior," "ruthlessly stratified characterization," and, most fundamentally, a "highly oversimplified depiction of human behavior."[25]

Janowitz points to the same tendency toward oversimplification and exaggerated claims for prediction and control in his contrast between research based on what he calls the engineering model and research based on what he calls the enlightenment model.

The engineering model emphasizes precision and the search for causality. Unfortunately (for the engineering model), changes in the present are constantly altering the emerging future. These changes thwart the efforts of the social engineer. The enlightenment model seeks to fuse research about the past with research of the present for the purpose of stimulating creativity, problem solving, and institution building.[26]

Janowitz's point is that, though its claims are more modest, enlightenment-model research actually produces a much greater benefit than engineering-model research in guiding the constructive adaptation of social organizations, including the military.

Alisdair MacIntyre makes the point more generally and even more strongly:

These two characteristics, total or near total predictability on the one hand and organizational effectiveness on the other, turn out on the basis of the best empirical studies we have to be incompatible . . . An effective organization has to be able to tolerate a high degree of unpredictability within itself.[27]

Accordingly, the notion that modern management techniques can actually deliver the results promised through precise planning and control is found to be a myth: " 'Managerial effectiveness' . . . is the name of a fictitious, but believed in reality, appeal to which disguises certain other realities; its effective use is expressive."[28]

Social science has a long history of analysis of institutions and organizations that emphasizes the limits of prediction and control. William Graham Sumner introduced the term *crescive institution* to denote the process of unanticipated, evolutionary adaptation.[29] Alfred Chandler's historical analysis of major business corporations showed clearly how effective structural forms developed painfully through trial and error, and it demonstrated that even the most intelligent and visionary business leaders often grope in the dark for successful adaptations to changing circumstances.[30] James Thompson's concept of bounded rationality and his exegetical treatment of organizational process as the emergence of mechanisms for coping with complexities and uncertainties that cannot be eliminated serve as effective antidotes to the scientific management precepts of Frederick Taylor and his associates.[31] Yet, the tendency of bureaucratic rationalism to overreach its capacities persists and thrives. As Shils points out, "Rationalism has thus far been successful because it has not been completely successful. The more successful it becomes the more it endangers itself, the more it lays itself open to resistance."[32]

We will return to this point of the self-generating dynamics of resistance to rationalism in the particular case of the military. However, in the short-term here and now, rationalism and the myth of managerial effectiveness often outmuscle whatever countervailing forces might exist. This suggests the operation of a type of Gresham's Law: the engineering model drives out the enlightenment model, economics drives out sociology (i.e., real sociology, not watered-down economics), arrogance drives out humility. An illuminating exercise would be, for example, to distribute the Department of Defense's total expenditures on personnel-related research between the engineering model and the enlightenment model. Resources tend to flow toward those who make the strongest promises of performance, which helps to account for the imitative response of other social sciences to the rise of economics.

The Consequences of Bureaucratic Rationalism for Military Organization: Recruitment

The consequences of the increased influence of bureaucratic rationalism for the military organization can be put into two categories using the lexicon of system analysis: inputs and outputs. The relevant input for the military in the context of the I/O thesis is personnel recruitment.

The effects of the dominance of bureaucratic rationalism on recruitment into the all-volunteer force can be examined again from the perspective of two categories: the characteristics of those who are recruited and the characteristics of those who are not. First, when appeals to recruitment and the whole strategy of personnel procurement are based on the marketplace mode of bureaucratic rationalism, some recruits (not all) enlist for reasons that turn out to be incompatible with the actual nature of military service. The result is attrition at remarkably high levels, concentrated primarily among those who have enlisted in direct response to market-oriented inducements rather than in response to other sources of motivation, such as patriotism or family tradition. The calculation of pay elasticities notwithstanding, the military probably cannot compete effectively in the labor market with the civilian sector if the former must compete exclusively on the latter's terms.

Fortunately for the ability of the all-volunteer force to meet its staffing goals, at least most of the time, the majority of recruits have enlisted with at least some of their motivation being noneconomic.[33] A reasonable estimate would be that only 10 percent to 20 percent of all recruits at any one time have enlisted for the "wrong reasons" (i.e., to get a job, any job, etc.); most of them exit through attrition, after having drained the resources of the military through the costs of recruiting, equipping, training, and administering discharge procedures. Some of these recruits would have enlisted under any conditions, but their numbers are probably magnified by advertising and recruitment appeals that emphasize pay and bonuses.

The other side of the effect on recruitment is in the nature of those who are not recruited into the all-volunteer force. The middle- to upper-middle-class male whose grades qualify him for entrance into a four-year college is increasingly rare in the enlisted ranks. When recruitment appeals are essentially occupational, a young man is unlikely to enlist if doing so makes no sense in occupational terms. The army's "Be All That You Can Be" advertising slogan—the ultimate me-decade line—cannot be effective in recruiting those for whom the army clearly does not offer a plausible means for maximizing their potential.

The failure of the military to attract middle-class, college-eligible youth's as recruits may be a result of a self-fulfilling prophecy. This group was defined as outside the target market, as unlikely to respond to marketplace inducements offered by the all-volunteer recruiting strategy; therefore, little, if any, attention or effort was directed toward considering how its members might be effectively recruited. Those who plan military recruiting seemed to have overlooked the obvious fact that many of these people have been willing—sometimes all too willing—to subordinate self and self-interest to organizational purpose in fraternities, football teams, civilian conservation corps activities, fundamentalist churches, and religious, mystical, and ideological cults of all sorts.

The distortions in recruitment produced by overreliance on marketplace mechanisms have resulted in a serious question about the social representativeness of America's armed forces; in serious shortages of well-educated, high-aptitude personnel; and in a costly attrition, which not only drains resources but also inflicts a destabilizingly high rate of turnover.

The Consequences of Bureaucratic Rationalism for Military Organization: Effectiveness

One of the most striking features of bureaucratic rationalism is the tendency to neglect or ignore the question of outputs. The imbalanced emphasis on inputs reflects the social characteristics of bureaucratic rationalism, one of them being the preference for what appear to be precise measurements:

> Because of the difficulties in using and interpreting . . . output measures, analysts and planners have come to rely more on input oriented proxies such as the quantity and quality of new recruits. The rationale for this approach is twofold. First, even though the measures of inputs are far from perfect, they are better than the available measures of output. Second, it is easier to determine the extent to which the volunteer force is responsible for changes in manpower inputs than it is to determine the extent to which the volunteer force is responsible for changes in military capability.[34]

This tendency to prefer to measure what is easily measured and to ignore that which is more difficult to measure is carried over to what types of input measures are used. Thus, recruitment is typically assessed in terms of the average number of years of education (as well as AFQT scores) rather than the quality of educational preparation brought by recruits to the military, and the question of what kind of attitudes and motivations recruits enlist with is generally ignored by those who officially monitor the success of the all-volunteer force.

Second, and perhaps even more fundamental, the tendency to ignore the question of outputs is facilitated by the implicit assumption that the transformation of inputs to outputs is not essentially problematic. The ultimate effectiveness of the all-volunteer force can therefore be accurately predicted simply from the quality of its inputs.[35]

This predilection to assume that all we need to know about the all-volunteer force is the quality of personnel it attracts perhaps partly reflects the increasing rarity of the military's being actively called upon to do what it is designed to do. However, combat has not been abolished, and the all-volunteer force has been tested in various ways in Iran, Grenada, and Lebanon. Further, one certainly might argue that the all-volunteer force is called upon to perform its mission (of deterrence) every day, and that its effectiveness at this should not be mechanically extrapolated from any input measures, whether staffing statistics or funding levels.

Luttwak points to what he calls the materialist bias and the tendency to ignore key nonmaterial, intangible factors:

> Actually the displacement of "conflict" effectiveness by business efficiency is merely part of a much larger phenomenon—the pervasive materialistic bias that distorts our entire approach to defense policy and military matters in general. With few exceptions (as when nuclear weapons are at issue), Pentagon officials, military chiefs, Congress, and the media all focus their attention on the measurable, material "inputs" that go into the upkeep and growth of the armed forces—i.e., the weapons and supplies, maintenance and construction, salaries and benefits . . . But when it comes to military power, the relationship between

material inputs and desired outputs is not proportional; it is in fact very loose, because the making of military strength is dominated by nonmaterial, quite intangible human factors, from the quality of national military strategy to the fighting morale of individual servicemen.[36]

In addition to national strategy and morale, Luttwak's list of key intangible factors includes organizational quality, quality of operational methods and tactics, combat skills, leadership, and cohesion. The intellectual adversaries of narrow bureaucratic rationalism and disproportionate reliance on the engineering model have strenuously argued for the importance of measuring combat readiness and combat effectiveness, particularly in a posture of deterrence. As Janowitz notes:

> Both official and private research institutes concerned with military power focus on estimating the number and types of weapons and the size of the personnel of the various military establishments of the world. The collection and assessment of these statistical data is of crucial importance and is more feasible than probing the "tough" question of the combat readiness and effectiveness of these military forces. Under classical conditions, the pace of war making was such that there was time to improve military effectiveness in the period between the declaration of war and the actual use of military forces. But under a strategy of deterrence, the worth of a military unit rests on its existing and immediate combat effectiveness—both real and perceived. Moreover, the use of expanded all-volunteer forces in the West to implement the NATO goal of deterrence has given special importance to the assessment of combat readiness.[37]

Sorley takes the case one step further, arguing that what output measures are used are inadequate, inaccurate, and distorting:

> Combat readiness . . . has tended to be measured using essentially quantitative criteria. Not only have these criteria, and the institutionalized methods of their application, had dispiriting and thus undermining effects on what they purport to measure, but in failing to measure that meaningfully they have provided misleading indications which have further skewed the application of productive reform. Thus there have resulted numerous policies and practices, from assignments and tenure to selection for promotion and command, which tend to erode unit cohesion, and thus to interfere with the development of genuine combat readiness.[38]

Service people at all ranks can perceive the disparity among the criteria for promotion and advancement on the one hand, what is really important to the mission of the military on the other, and what is in fact measured.

Thus, we arrive at the hypothesis that the preeminence of bureaucratic rationalism in planning and managing the all-volunteer force has had serious, deleterious effects on the ability of the armed forces to perform their missions of deterrence and, as necessary, combat. The data are insufficient to test this hypothesis conclusively, partly because bureaucratic rationalism tends to ignore or give scant attention to the measurement of effectiveness and to crucial variables affecting readiness, such as morale, leadership, and organizational structure. What data are available, however, strongly suggest that the organizational effectiveness of the U.S. armed forces falls well short of what it might be and that the deficiencies can in large part be traced to policies and rhetoric derived from bureaucratic rationalism.

Since the inception of the all-volunteer force, a number of studies have produced alarming findings with regard to the attitudes and morale of enlisted personnel. A 1976 DOD survey of enlisted personnel discovered that a quarter of all E-4s (the modal rank of enlisted personnel) either disagreed with or were neutral toward the statement, "The mission of the military is both meaningful and important."[39] Wesbrook's study of junior enlisted men in combat arms units found that "28% believe that their senior NCO's and officers cannot be trusted, and 23% more are uncertain; 43% feel that they are accomplishing nothing as soldiers, and 19% more do not know if they are accomplishing anything or not."[40]

The purpose of this chapter is not to attempt to document exhaustively all the data bearing on the attitudes, morale, and cohesion of personnel serving in the all-volunteer force. Rather, I suggest the working hypothesis (well supported by the institution-occupation literature) that morale, cohesion, and attitudes toward military service and leadership fall well short of optimal levels. (The use of *optimal* suggests the interesting possibility of whether the military could ever be too institutional, an important question that is also beyond the scope of this chapter.)

A number of intervening variables may be seen as the specific conditions that derive from the overall influence of bureaucratic rationalism and that adversely affect morale, cohesion, and effectiveness. These factors include the dispiriting effects of inappropriate measures of effectiveness and readiness (often called the numbers game), excessive rotation of commanders, promotion criteria that reward low-risk management over risk-taking leadership, disproportionate reliance on monetary incentives in recruitment and retention, an individualized personnel system that tends to ignore group dynamics such as cohesion, ambiguities about the exposure of women to combat associated with the increased use of female personnel, insufficient training resources and realism in training, increased use of civilian contract labor derived from narrow cost-benefit calculations, and the changing composition of the enlisted force primarily reflecting the underrepresentation of white middle-class males.

It is difficult to determine the consequences of the diminished morale, cohesion, and leadership that follow from the preceding conditions and, ultimately, from the dominance of bureaucratic rationalism. Luttwak makes a strong case (not a conclusive case, which would be more than difficult) that the performance of elements of the all-volunteer force in Iran, Beirut, and even in the successful operation in Grenada was highly unsatisfactory.[41] His critique focuses on an analysis of the structural deficiencies of the U.S. armed forces: "At present, the defects of structure submerge or distort strategy and operational art, they outrightly suppress tactical ingenuity, and they displace the traditional insights and rules of military craft in favor of bureaucratic preferences, administrative convenience, and abstract notions of efficiency derived from the world of business management."[42] Thus, according to

Luttwak, excessive bureaucratic rationalism leads to inappropriate organizational structure and the neglect of the intangibles of morale, cohesion, and leadership and thus to operational ineffectiveness—in Vietnam and afterwards. In conclusion, then, our hypothesis is that the preeminence of bureaucratic rationalism (most prominent in the marketplace economics models of all-volunteer force planners but also evident in other areas) has at best placed organizational effectiveness of the military in serious question and at worst has severely undermined the military's capacity to perform effectively, both in deterrence and in combat.

Organizational Response to Bureaucratic Rationalism

We have previously noted Shil's observations that as bureaucratic rationalism becomes increasingly potent it generates increased resistance to itself. This is strikingly true in the case of the military, in which self-corrective dynamics have at least partly ameliorated the effects of bureaucratic rationalism.

First, there is a direct, often visceral, rejection of the occupational rhetoric of the management of the all-volunteer force. One of the most dramatic examples of this is a narration of a "soldier's story," produced by Col. Dandridge Malone and quoted in James Fallows's *National Defense*:

> Malone tells the soldier's story, from the time he leaves home, a young recruit, on his way to boot camp . . . the anxiety and confusion at the training schools; the friendships, the coarseness, the constant reassignments and promotions; the compromises and satisfactions of military marriage; and on to Vietnam, through the fire fights, the fear again, the deaths of friends; survival and return; the first glimpse of children he has not seen for a year, the first embrace of his wife . . . "—and if all these wondrous things," Malone draws at the end . . . "which thousands of us share in whole or part, can, by the mindless logic of a soulless computer, programmed by a witless pissant ignorant of affect, be called, *just another job* . . . then, by God, I'm a sorry, suck-egg mule."[43]

Even the widespread expressions of interest during the mid-1970s in the possibility of joining a military union could be seen as a reaction against rather than an acceptance of an occupational definition of the situation. Some of the most militantly pro-union NCOs were outraged that, in their view, Congress was treating them as labor to be bought and sold according to prevailing prices.

Second, the elements of tradition, patriotism, and institutional values are deeply rooted in the armed forces and society. Market-model policies, advertising, and recruiting cannot overcome the power of mass media socialization (war movies, television programs, popular books, and even cartoons and comics), the intergenerational linkages of the military service of fathers, uncles, sons, and nephews (and mothers, aunts, daughters, and nieces); the potent symbolism of the flag, the uniform, weapons, and ceremonies; and, of course, the reality of risk, valor, and death in military operations reported and graphically portrayed on television. A review of data bearing on enlistment decisions indicated that approximately 80 percent or more of recruits were motivated by patriotic sentiments (as well as other factors), despite the absence of patriotic appeal in recruiting strategies.[44]

Third, certain institutional responses—for example, Project Warrior in the air force, General Meyer's initiatives to improve cohesion in the army—have been designated and implemented deliberately to offset or counter the deleterious consequences of bureaucratic rationalism. Some researchers, notably Larry Ingraham and Frederick Manning, have carefully and illuminatingly examined the intangibles of morale, cohesion, and leadership.[45]

However, the perspective of bureaucratic rationalism is resilient. One of the foremost apologists for the rationalist point of view, Richard Hunter, brushed aside any concern about intangibles by asserting that "peacetime soldiers, and especially garrison soldiers, always suffer from morale, motivational, and attitudinal problems that are best solved by exposing them to the combat stimulus and permitting them to do what they have trained to do in a situation where it really counts."[46] Turning to the words of the senior DOD manpower official, Lawrence Korb, is more heartening: "If the American military intends to maintain the All-Volunteer Force which accepts 'marketplace' values and contractual relationships, military professionals must take extra care to inculcate their troops with the virtues which transcend the values upon which their contracts are based, and stress the notion of duty for its own sake."[47] This sentence appears in a book published in 1981, and one only wonders how much energy within the Office of the Assistant Secretary for Manpower has been devoted to the calculation and management of contractual "marketplace" relationships and how much toward the inculcation of transcendent values.

DISCUSSION AND METHODOLOGICAL NOTE

One of the central conclusions here is that Moskos's original formulation of the I/O thesis is correct: the master variable is a shift in the rationale of the military (flowing from bureaucratic rationalism), and from this shifting rationale come changes in both structure and function.

Without question, important changes in the structural organization of the armed forces have resulted from this changing rationale; also without question, these transformations in structure have affected the attitudes and performance of military personnel. However, more than one structural route leads to organizational effectiveness. Suggesting that an effective armed force is incompatible with off-post housing of junior enlisted personnel is as narrow-minded as nineteenth-century naval officers' prediction of the collapse of discipline and willingness to fight with the abolition of flogging.

The single greatest structural change in the post-Vietnam military is the termination of conscription. The shift to the all-volunteer force is rightly seen as a major point of inflection in the pattern of organizational change identified by the I/O terminology. We must recognize, however, that this increased impetus toward bureaucratic rationalism does not inhere in the change from conscription to all-volunteer recruitment. Rather, it results from the way the change was implemented. The shift away from a draft system that had become morally debased to an all-volunteer force presented a major opportunity to reconstruct the meaning of military service within the context of intelligent and responsible citizenship. That opportunity was missed, but today is not too late to develop programs of civic education and inspiration that reinforce and strengthen the institutional values of military service, both in the armed forces and civilian society.

Of course, additional empirical research on the I/O model is needed. Most of the research on I/O trends and strains in the military has relied on surveys (one of the preferred social research tools of bureaucratic rationalism, incidentally), with the focus on the attitudes of service members toward the military and toward their "jobs." Moskos has suggested using a set of attitudinal questionnaire items to explore the subject more systematically. These might include, for example, the following positions:

> What military members do in their off-duty hours is none of the military's business.
> Compensation should be based primarily on one's merit and not on rank and seniority.
> The military requires participation in too many activities that are not related to my job.
> If I had an independent income of $75,000 a year, I would still stay in the military.[48]

Most of Moskos's items, once again, are best suited for measuring what I have called occupationalism.

Because measuring structural changes in the military in addition to the social psychology of occupationalism is important, I would also suggest the need for another set of attitudinal items designed to measure the degree and attachment to bureaucratic rationalism on the part of those who shape the military structure. These items might include the following:

> It is possible to control the direction of complex organizations through modern management techniques.
> With the proper data and analytical methods, one can accurately explain, predict, and control manpower processes such as recruitment and retention.
> Marketplace incentives and econometric analysis are the best tools for managing the recruitment of personnel to the military.
> Social institutions are the product of long-term processes of

development and therefore often involve latent functions that can be unnoticed and unappreciated by senior officials.

Organizations, and those who attempt to manage them, must cope with extensive uncertainty on multiple fronts.

Deliberate attempts to bring about change within an institution often triggers unanticipated (and sometimes undesired) consequences.

Efforts to understand human behavior exclusively in terms of self-interest will inevitably be confounded by reality.

I would also underscore the importance of breaking past the tendency to restrict the research enterprise to large-scale surveys and thereby including the full battery of social scientific techniques—historical analysis, anthropological field studies, life history techniques, and experiments (especially field experiments).

The foregoing exercise is an attempt at conceptual clarification. Clearly, the trends in military organization associated with the I/O thesis are dangerous for both military effectiveness and civil-military relations. Those responsible for these areas must be able to understand clearly the nature of the dynamics. The argument here is that this understanding is advanced when we differentiate the attitudes of military members (the issue of occupationalism) from the orientation of officials who manage the military (the issue of bureaucratic rationalism) and when we see that the latter is related to the former as a crucial causal factor, operating both directly and through the intervening variables of structural change.

NOTES

1. Charles C. Moskos, "From Institution to Occupation: Trends in Military Organization," *Armed Forces & Society* 4 (1977): 41–50.
2. Morris Janowitz, 'From Institutional to Occupational: The Need for Conceptual Clarity," *Armed Forces & Society* 4 (1977): 51–54.
3. Readers familiar with the history of sociology will note the parallel between Moskos's use of *institution* and *occupation*, and Tonnies's *Gemeinschaft* and *Gesellschaft*, Durkheim's *organic solidarity* and *mechanical solidarity*, and Park's *community* and *society*. See Robert E. L. Faris, *Chicago Sociology 1920–1932* (San Francisco, Calif.: Chandler, 1967).
4. Moskos, "From Institution to Occupation," p. 45, emphasis added.
5. Zig Ziglar, *Zig Ziglar's Secrets of Closing the Sale* (Old Tappan, N.J.: Fleming H. Revell Co., 1984), pp. 130–131.
6. Donald F. Roy, "Work Satisfaction and Social Reward in Quota Achievement," *American Sociological Review* 18 (1953): 507–14; Donald F. Roy, "Quota Restriction and Goldbricking in a Machine Shop," *American Journal of Sociology* 62 (1952): 427–442.
7. Christopher W.L. Hart and Gregory D. Casserly, "Quality: A Brand-New, Time Tested Strategy," *The Cornell Hotel and Restaurant Quarterly* 26 (November 1985): 52–53.
8. Richard T. Pascale and Anthony G. Athos, *The Art of Japanese Management: Applications for American Executives* (New York: Simon and Schuster, 1981).
9. William Ouichi, *Theory Z* (Reading, Mass.: Addison-Wesley, 1981).
10. Thomas J. Peters and Robert H. Waterman, Jr.' *In Search of Excellence* (New York: Harper and Row, 1982).
11. Ibid., p. 29.

12. Tom Peters and Nancy Austin, *A Passion for Excellence* (New York: Random House, 1985), p. xix.
13. A careful reading of Homer makes clear that what angered Achilles was not the loss of goods so much as the symbolism of being undervalued.
14. Morris Janowitz, *The Professional Soldier* (New York: Free Press, 1960), pp. 104, 117.
15. Jonathan Alford, "Deterrence and Disuse," *Armed Forces & Society* 6 (Winter 1980): 247–256.
16. Robert H. Baldwin and Thomas V. Daula, "The Cost of High-Quality Recruits," *Armed Forces & Society* 11 (Fall 1984): 111; Aline O. Quester and James S. Thomason, "Keeping the Force: Retaining Military Careerists," *Armed Forces & Society* (Fall 1984): 93.
17. Robert F. McNown, Bernard Udis, and Colin Ash, "Economic Analysis of the All-Volunteer Force," *Armed Forces & Society* 7 (Fall 1980): 128–129.
18. Charles C. Moskos, Jr., "Serving in the Ranks: Citizenship and the All-Volunteer Force," as cited in McNown, Udis, and Ash, "Economic Analysis," p. 113.
19. Sheldon S. Wolin, "The New Public Philosophy," *Democracy* (October 1981): 28.
20. Lester C. Thurow, *Dangerous Currents: The State of Economics* New York: (Random House, 1983).
21. See Charles C. Moskos's book review in *Contemporary Sociology*, 13, no. 4 (July 1984): 420–421.
22. Irving Kristol, "The Adversary Culture of Intellectuals," in Seymour Martin Lipset, ed., *The Third Century* (Stanford: The Hoover Institution Press, 1979), pp. 328–343.
23. Edward Shils, *Tradition* (Chicago: University of Chicago Press, 1981), p. 7.
24. John Keegan, *The Face of Battle* (New York: Viking Press, 1976), p. 36.
25. Ibid, p. 39.
26. Morris Janowitz, "Consequences of Social Research on the U.S. Military," *Armed Forces & Society* 8 (Summer 1982): 522.
27. Alisdair MacIntyre, *After Virtue: A Study in Moral Theory* (Notre Dame: University of Notre Dame Press), p. 100.
28. Ibid., p. 73.
29. William G. Sumner, *The Folkways* (Boston: Ginn and Co., 1906).
30. Alfred D. Chandler, *Strategy and Structure* (Cambridge, Mass.: MIT Press, 1962).
31. James D. Thompson, *Organizations in Action* (New York: McGraw-Hill, 1967).
32. Shils, *Tradition*, p. 316.
33. James S. Burk with John H. Faris, "The Persistence and Importance of Patriotism in the All-Volunteer Force." Report to U.S. Army Recruiting Command by Battelle Memorial Institute, 1982.
34. Richard V.L. Cooper, "The All-Volunteer Force: Status and Prospects of the Active Forces," in A.J. Goodpaster, L.H. Elliot, and J.A. Hovey, Jr., eds., *Towards A Consensus on Military Service: Report of the Atlantic Council's Working Group on Military Service* (Elmsford, New York: Pergamon Press, 1982), pp. 76–113.
35. See, for example, Charles Dale and Curtis Gilroy, "Determinants of Enlistments: A Macroeconomic Time-Series View," *Armed Forces & Society* 10 (Winter 1984): 204.
36. Edward N. Luttwak, *The Pentagon and the Art of War: The Question of Military Reform* (New York: Simon and Schuster, 1984), p. 139.
37. Morris Janowitz, "Editor's Note: A Special Issue on Combat Readiness for a Deterrent Strategy," *Armed forces & Society* 6 (Winter 1980): 169.
38. Lewis Sorley, "Prevailing Criteria: A Critique," in Sam C. Sarkesian, ed., *Combat Effectiveness* (Beverly Hills, Calif.: Sage, 1960), pp. 57–93.
39. John H. Faris, "Economic and Noneconomic Factors of Personnel Recruitment and Retention in the AVF," *Armed Forces & Society* 10 (Winter 1984): 261.
40. Stephen D. Wesbrook, "Sociopolitical Alienation and Military Efficiency," *Armed Forces & Society* 6 (Winter 1980): 178.
41. Luttwak, *Pentagon and the Art of War*.
42. Ibid., p. 65.
43. James Fallows, *National Defense* (New York: Vintage Books, 1981), pp. 108–109.
44. Burk, "Persistence and Importance of Patriotism."
45. See Larry H. Ingraham, *The Boys in the Barracks: Observations on American Military Life* (Philadelphia: Institute for the Study of Human Issues, 1984).
46. Richard W. Hunter, "An Analysis of the All-Volunteer Armed Forces—Past and Future," in Robert K. Fullinwider, ed., *Conscripts and Volunteers: Military Requirements, Social Justice and the All-Volunteer Force* (Totowa, N.J.: Rowman and Allenheld, 1983), pp. 23–45.

47. Lawrence J. Korb, "Future Challenges," in James H. Buck and Lawrence J. Korb, eds., *Military Leadership* (Beverly Hills, Calif.: Sage, 1981), pp. 235–241.

48. Questionnaire items distributed at the International Conference on Institutional and Occupational Trends in Military Organization at the U.S. Air Force Academy, Colorado Springs, June 12–15, 1985.

Part Two

Issues in the American Military

VI

The Military and the Family as Greedy Institutions

MADY WECHSLER SEGAL

The study of military families involves analysis of how two societal institutions—the military and the family—intersect. Both make great demands of individuals in terms of commitments, loyalty, time, and energy: they therefore have many of the characteristics of what Lewis Coser calls "greedy" institutions.[1]

My contention is that, due to various social trends in American society and in military family patterns, there is greater conflict now than in the past between these two greedy institutions. Further, the ways in which they respond to this competition for the service member's commitment are already affecting and will continue to affect how far the military moves in an institutional or occupational direction. Moreover, changes can occur within the traditional military institution without the military becoming less institutional and more occupational.

THE NATURE OF GREEDY INSTITUTIONS

Both the military and the family depend for their survival on the commitment of the members who have dual loyalties to the military and to the family. Coser notes that individuals can meet competing demands because "modern social institutions tend to make only limited demands on the person."[2] He illustrates this with the example of how the demands of work and family can be reconciled: "The amount of time that an individual legitimately owes to his employer is normatively and even legally established; this makes it possible for him to have time for this family or other non-occupational associations."[3]

These normative and legal limitations on employers' demands are relatively recent. Historically, employers often had greater control over their employees' time. This was certainly true under feudal systems, but it also characterized systems of apprenticeship. Even after industrialization, prior to the trade union movement and legislative restrictions, factory owners controlled the working conditions of their employees; they could and did require long hours. The company town was distinguished by one employer not only controlling the work lives of residents, but also exercising greater control over the nonworking hours of employees. The trade union movement brought about guarantees of better working conditions and legislative restrictions on the working hours employers could exact from employees. In contrast, unionization of the American military has been statutorily precluded.

Although the military and the family may not possess in extreme form all the characteristics Coser identifies for greedy institutions, they fit the concept well enough to make it useful in analyzing military families. I contend that the current competition between the military organization and the family is occurring in a period of social change without an established normative pattern and that it will lead to new normative patterns for resolving the conflicts. The potential outcomes of this institutional adaptation process can be analyzed in terms of the institutional-occupational (I/O) model of military organization by exploring whether they move the military organization in a more institutional or occupational direction. We must recognize, however, that the change may be from one institutional form to another.

For institutions, and groups and organizations within them, to survive in the face of competing demands on individuals, they must develop mechanisms for motivating individual participation and commitment. In the institutional military, individual commitment and self-sacrifice is legitimated through the operation of normative values, which compel the individual to accept great demands on his time and energy. Further, the organization controls the demands: the individual does not get to choose when and how to comply. Role obligations are diffuse, and place of residence is not separated from place of work. In return for his service, the individual receives esteem from the larger society and compensation from the military, much of it in noncash form. Of the several types of nonmonetary compensation, those that have the greatest relationship to families are job security, on-base housing, medical care, allotments by family size, subsidized on-base consumer facilities, and numerous on-base services, such as schools and recreational activities.

To accomplish its mission, the military makes various demands on service members. Although it exerts some specific normative pressures directly on family members, most pressures affecting families are exerted indirectly through claims made on the service members. For both types of pressures, the family is expected to adapt to the greediness of the military institution and support the service member in meeting military obligations. However, important societal trends in general and in military family patterns in particular are making this adaptability problematic. Because of these trends, which include changes in women's roles in society (especially labor force participation rates), as well as increases in the numbers of married junior enlisted personnel, sole parents, active-duty mothers, and dual-service couples, military families themselves are becoming greedier, increasing the potential conflict between the military and the family.

To clarify the conflict, the ways in which the family operates as a greedy institution, especially for certain people, will be examined. Second, the military will be analyzed as a greedy institution; the specific demands it makes—and some effects those demands have—on service members and their families will be described. Third, the trends that lead to increased conflict will be detailed. Further, actual and potential adaptations of the military to family trends will be discussed.

THE FAMILY AS A GREEDY INSTITUTION

Nuclear families make different demands on different members. All members are expected to be emotionaly committed to the family, to display affection toward other members, to identify with the family as a unit, and to fulfill role obligations that are diffuse, relative to most other social groups. However, the family is not a greedy institution for all members; rather, if we use the concept of greed as a continuous dimension, the family is greedier for some members than for others.

Lewis Coser and Rose Coser note that the family is greedy for women and discuss recent changes in the normative legitimacy of this greed. They find that not the husband but the wife "is expected to devote most of her time, as well as her emotional energies, to their family."[4] They discuss how housewives are expected to maintain exclusive attachment to the family and to make sacrifices for it. For example, as both cause and consequence of the greediness of the family, women have been excluded from high-status occupations because such occupations require commitments that would interfere with the fulfillment of family obligations.

The insights Coser and Coser provide into the dynamics of the conflict between family demands and occupational requirements for professional women have broad applicability. Although this conflict is generally greater for women, it occurs for men as well, to the extent that their work interferes with their ability to provide the normatively expected resources of time, energy, and affect to the family. This happens when either the family is relatively greedy or the work itself is so greedy that it interferes with the fulfillment of relatively minimal family demands.

At certain stages, of course, the family is relatively greedy for both women and men. This is the case, for example, with new marriages, which require greater time and emotional adjustment than established relationships, and with children who are very young or who for other reasons need fairly constant supervision. As children get older, they need less constant supervision; however, parents of adolescents often find this stage demanding emotional energy. Clearly, the family is especially greedy when it has only one parent.

Moreover, although care of children and household tasks have been traditionally considered women's work, fathers never have been absolved from all family responsibilities, and social changes are increasing their normative share. Witness how the housewife role is becoming a much less exclusive role for women. The past 20 to 30 years have witnessed a virtual revolution in women's labor force participation; today, a majority of American women, including those with preschool children, are in the labor force. Legislation has removed many educational and occupational barriers to women, increasing their access to high-status positions.

Looking toward the future, Coser and Coser envision a time when, as a result of both more flexible work schedules and decreased household tasks, the pattern of women's work and family roles will resemble men's. The Cosers' vision includes greater continuity between women's lives before and during marriage, which they conclude "will be the death of the 'greedy' family."[5] Although I agree that the family of the future will cease to be greedy in the extreme for women and that overall family demands will decrease, I contend that the more wives resist the greediness of the family and participate in the work world, the greater will be the family demands on men. This increases the potential for conflict not only between husbands and wives, especially during the transition to greater equality between men and women at home and at work, but also between work and family demands for men, especially for those in greedy occupations such as the military. For instance, we can expect pressures from wives on husbands to adapt their career decisions to family needs, including wives' career considerations.

THE MILITARY AS A GREEDY INSTITUTION

The military is unusual in the pattern of demands it makes on service members and their families. Although each specific organizational requirement can be found in other occupations, the military is almost unique in the constellation of requirements. (Perhaps the only other occupation that exerts a similar set of pressures is the foreign service.) Some demands vary in frequency and intensity among and within the services, but over the course of a military career, a family can expect to experience all the specific demands. Characteristics of the life-style include risk of injury or death of the service member, geographic mobility, periodic separation of the service member from the rest of the family, and residence in foreign countries. Normative pressures are also directly exerted on family members regarding their roles in the military community.

Risk of Injury or Death

The risk that military personnel will be wounded or killed in the course of their duties is an obvious aspect of the institution's demands. The legitimacy for the institution to place its members at such physical risk is perhaps the greediest aspect of all. Although this risk is, of course, greatest in wartime, even peacetime military training maneuvers entail risk of injury, and military personnel can be sent at any moment to areas of armed conflict. The effects on military families of the potential for injury and death in both peacetime and wartime are studied relatively seldom. Therefore, we know less about them than about the effects of other military demands on families. The potential for casualty has been studied as a source of stress during certain kinds of separations. Studies of military families conducted during and after World War II emphasized the impact of wartime separation. Similarly, recent research has focused on the families of American prisoners of war and those missing in action in Vietnam. There is much less research on the grieving process in families of servicemen who were killed.[6]

Geographic Mobility

The U.S. armed forces periodically transfer personnel to new locations, for new tours of duty as well as for specific types of training. Both frequency of mobility and length of duty tours vary among the services as a function or organizational policy. Results of the 1978–79 DOD Survey of Officers and Enlisted Personnel show that, across all services.

> [twenty-nine] percent of surveyed enlisted personnel have been at their present location less than one year, 36 percent between one and two years, and 21 percent between two and three years. Among officers, 33 percent have been at their present location less than one year. 33 percent between one and two years, and 22 percent between two and three years.

Thus, 86 percent of enlisted personnel and 88 percent of officers had moved at least once in the three years preceding the survey.

Families of military personnel move less often than the service members themselves, but relocation is still frequent, especially for officers' families. For all personnel, regardless of length of service, the number reporting at least three family moves was 43 percent for enlisted personnel and 69 percent for officers; for those with seven to 10 years of service, the corresponding figures are 41 percent and 59 percent, respectively. For enlisted personnel who had been in the military for more than 14 years, 9 percent reported that their spouse and/or children had moved more than nine times; the corresponding figure for officers is 31 percent.[8]

Although many service members and their spouses view the opportunity to travel as a benefit of military service, geographic mobility is also seen as a hardship that disrupts family life and necessitates adjustments under the best of circumstances.[9] The most obvious component of geographical mobility is the requirement to move frequently. Less obvious and less often discussed in the literature on military families is the first move, which perhaps calls for the greatest adjustment. Whether a service member is already married when entering the military or marries later, the spouse's first residence during military service is usually away from home. This has special implications for young enlisted families because they are likely to be away from their families of origin and long-term friends for the first time. In contrast, junior officers are older and have gone to college, often away from home. Similarly, officers' spouses are more likely than enlisted spouses to have gone to college, though they too (like enlisted spouses) may be away from their extended families for the first time. All, however, are geographically separated from their usual interpersonal networks and sources of social support. If they manage to become integrated into new supportive networks, these relationships are severed when they are required to move to a new location, an experience that is repeated with each additional move.

Besides the social disruption of moving are the other adjustments to a new location. The place is unfamiliar and one must learn one's way around. For families with limited resources and/or small children, this can be especially stressful. Junior enlisted families often must make do without a car or access to public transportation. The physical environment and climate may also be quite different from what the family members are accustomed to. Whatever the type (including size) of the community, some military families will be used to another type. For example, people from rural areas often find large military installations intimidating; they are unaccustomed to living close to neighbors. The regional dialect may also be unfamiliar to them, straining communication with others and possibly making them feel like outsiders.

Moving affects families differently, depending on where they are in the life cycle. Those early in their military careers (especially enlisted personnel) have the least control over when and where they move. They also have the fewest military institutional supports, such as family housing on post. For children in school, moving disrupts their education; the lack of standardized curricula among states can cause gaps or repetition in what they are taught. Moving can be particularly stressful for teenagers.[10] Adolescence involves a search for personal identity, which usually requires integration into a peer group; moving not only disrupts relationships with peers, but also hampers participation in extracurricular activities, which may be an important component of the teenager's self-concept.

Moving is especially harmful to spouses' employment possibilities and career continuity. Unemployment rates are substantially higher for military than civilian wives.[11] Family income is lower in military families than in civilian families, due largely to lack of employment opportunities for wives. For officer's wives, who are more likely to be oriented to a career rather than just to employment, geographic mobility interferes with normal career progression. Even if they can find work in their field, they lose seniority when they move. Thus, employment problems create economic hardships for the family and problems of personal identity and worth for the wives.

While the military provides subsidies for moving expenses when a service member is transferred to a new location, these are often inadequate, forcing the family to pay some of the expenses. For both officer and enlisted families, the frequent moves leave them less likely than civilian families to own their own home, which is not only a cultural goal in American society, but also is often a family's major financial asset.

Separation

Military demands often necessitate that service members be away from their families. The extent and nature of these requirements vary among and within the services. In peacetime, the most common types of assignments that result in family separations include military schooling, field training, sea duty, and unaccompanied tours. On some assignments, the family is allowed to accompany the service member but for various reasons chooses not to. The length of peacetime separations generally varies from a few days to a year; wartime separations can be much longer and indefinite. Certain military units (e.g., submariners and the 82d Airborne Division) experience frequent or prolonged separations.

Hill reports that 75 percent of families of career military personnel "have experienced one or more prolonged periods of father absence."[12] At the time of the 1978–79 DOD survey, 15 percent of all married enlisted personnel and 5 percent of officers reported that they were not accompanied by their families. Among the services, separation rates were lowest in the air force (8 percent of enlisted, 4 percent of officers) and highest in the Marine Corps (23 percent of enlisted, 10 percent of officers).[13]

These proportions of unaccompanied personnel do not include separations caused by temporary duty and field training. Thus, at any one time, the proportions of service members who are away from their families are much higher. Results of the 1978–79 DOD survey show that, of those with spouses and/or children, 55 percent of enlisted personnel and 63 percent of officers had been separated from their families for some time during the year preceding the survey, including 25 percent of enlisted personnel and 29 percent of officers who had spent at least five months away from their families.

Some effects of separations on families depend on the type of separation, and others are common to most, if not all. Separations always require adjustments by service members and their spouses and children. Because little attention has been given to service members either in research or in the provision of services, we know less about their reactions to separations than about those of their wives. The most common problems experienced by the latter are loneliness, problems with children, physical illness, and loss of their usual social role in the community. The wife is thrust into the role of sole parent. Research on military children with psychological problems has often identified the absence of the father as a contributing factor. Even when families cope well with separation, they still view it as a stressful experience requiring adjustment.[14]

The strains of separation may be especially difficult at certain stages of family life. Newly married couples, who have had less time to solidify their relationships, are vulnerable to the disruption of separation. Men who are away during their wives' pregnancies and/or childbirths miss an important family event. Service members separated from their young children miss periods of rapid growth and change in their children. Similarly, they sometimes are absent for special events in their children's lives (e.g., a child's first step, first word, first ride on a bicycle, birthday; an adolescent's participation in competitive athletic events, graduation from high school). Family separation during children's adolescence may also interfere with parent-child relationships and, therefore, with the adolescents' psychological development. In general, such absences can cause distress for the service family; sometimes family members harbor long-term feelings of resentment and abandonment.

Separations also have potentially beneficial effects. They allow for individual growth and for development of the marital relationship. Some relationships benefit from a period of less intense interaction. Military members and their spouses sometimes say a separation makes them appreciate each other more and adds novelty and romance to their relationships. Nevertheless, the difficulties of separation are usually seen to outweigh the benefits.

Residence in Foreign Countries

Because of the United States' role in the balance of world power, even in peacetime substantial proportions of American military personnel are stationed overseas. The numbers of troops in particular places vary as a function of U.S. foreign policy and world events. As of March 31, 1984, 39 percent of all military personnel were stationed outside of the continental United States; 25 percent were stationed outside of the U.S. states and territories. The corresponding figure for the army were 39 percent outside the continental United States and 34 percent outside the U.S. states and territories. From the 1978–79 DOD survey, Doering and Hutzler report the number of years military personnel had spent in overseas duty according to length of service. Across all services, 59 percent of enlisted personnel and 56 percent of officers with one to six years of service had spent at least one year overseas. For those with seven to 10 years of service, the corresponding figures are 90 percent and 80 percent, respectively.[15]

Periodic foreign assignments bring to the military family both benefits and hardships. Although relocation adjustments vary according to how different the host culture is from American society, even in those most similar to the United States (industrialized Western democracies), the initial reaction is often one of culture shock. Behavioral norms differ on matters both serious and mundane; language barriers can lead to feelings of isolation and even fear. Although most families eventually cope with living abroad—and many thoroughly enjoy the experience—some encounter severe difficulties. Economic problems are prevalent when the foreign currency exchange rate is unfavorable to the U.S. dollar. Spouses have difficulty obtaining employment. Families who are not command-sponsored—that is, those who are not provided with such support as transportation to the location for family members or housing on post (mostly junior enlisted families)—must live "on the economy" and tend to be isolated from formal and informal military institutional supports. Even for eligible families, on-post housing may not be available at the beginning of the service member's tour of duty. Thus, the family must either make two moves (the first to off-post housing) or experience family separation; with the latter option, adjustment to the foreign culture is coupled with reunion adjustments.

Normative Constraints

Although the demands on military service members impact indirectly on family life, the military also exercises more direct constraints on families through normative pressures on the behavior of spouses and children. Family members informally carry the rank of the service member, and behavioral prescriptions vary accordingly. In the traditional institutional military, wives of service personnel "are expected to initiate and take part in a panoply of social functions and volunteer activities."[16] Wives of officers and senior noncommissioned officers are integrated into a military social network with clearly defined role obligations and benefits determined by their husbands' ranks and positions.

Wives are socialized through various mechanisms. Traditional expectations are clearly spelled out in handbooks, such as Nancy Shea's *The Army Wife*, which cover military customs, rank courtesy, entertaining, etiquette, calling cards, and so on.[17] Wives are informed that their husbands owe their primary loyalty to the military, not the family:

> Early in your new role as an Army wife you must understand that your husband's "duty" will come first—before you, before your children, before his parents, and before his personal desires and ambitions.[18]

Additionally, various social obligations are specified:

> It is every wife's duty not only to join, but to take an active interest in the wives' club on the post where her husband is stationed.[19]

Although some of these prescriptions do not now carry the normative force they have had traditionally, most are still communicated and enforced through informal interpersonal processes.

Family members learn that their behavior is under scrutiny and that the degree to which it conforms to normative prescriptions can affect the service member's career advancement. Social pressures are especially exerted on officer's wives and, to a lesser extent, on senior NCO wives. For enlisted wives and for military children, there are fewer prescriptive obligations; rather, pressures on them are more proscriptive in nature. Enlisted wives are generally not required to engage in military community activities; they are expected, however, to refrain from "troublesome" behavior. Violations of such norms by enlisted wives or by children that come to the attention of military authorities result in pressure on service members to control their families. Such pressures are most likely for families who live on military installations because they are more subject to military control and attention than are those who live off post.

Although military wives may experience normative constraints as pressures, wives also benefit when they are incorporated into the military system. With their roles institutionalized, wives have defined social identities and are more readily integrated into supportive social networks. Such integrative social mechanisms are likely to make important contributions to personal well-being, especially during stressful times such as routine family separations, relocation, and combat deployment.

MILITARY FAMILY PATTERNS

The demands of the military organization (discussed previously) require certain kinds of adaptation from military families. Spouses and children must be willing to move when the service member is transferred. Families must be willing to endure separations; spouses must be willing, available, and able to perform the family roles vacated by the absent service member. Military social and volunteer activities depend on the participation of officers' wives.

Conflict between military requirements and family needs is avoided when the family adapts to the military's demands. However, family patterns of U.S. military personnel have changed dramatically over the past century; the past 15 years have seen further changes. Most notable (in that it characterizes the largest number of personnel) has been the increase in the proportion of military men who are married, especially in the enlisted ranks. The proportion of active-duty women has increased substantially; those who are married to military men enlarge the number of dual-service couples. The number of sole parents, that is, currently unmarried men and women with minor children, has also grown. Further, civilian wives of military men are being greatly affected by changes in the roles of women in American society that include, most notably, personal identities increasingly independent of husbands and higher rates of labor force participation. All these trends have meant a larger proportion of personnel for whom the family is greedy and, by normative definitions, legitimately so. This, in turn, means a heightened potential for conflict between the military and the family.

Marriage

During the late nineteenth and first half of the twentieth centuries, the vast majority of military personnel were unmarried. Marriage was the rule only among older officers; indeed, prior to World War II, *military wife* was virtually synonymous with *officer's wife*, because few enlisted men married and, if married, their families were of little concern to the military institution.[20] The proportion of married military men climbed steadily after World War II. From 1953 to 1960, the number of servicemen who were married rose from 38 to 52 percent; by 1980, 61 percent of all military personnel were married.

Thus, the American military has increasingly become a married military. This trend, which embraces majorities of the middle and upper ranks and substantial minorities of even the lowest ranks, represents a highly significant change in the nature of enlisted life in the armed services. Because married enlisted personnel do not live in the barracks, barracks life has lost some of its character as a total institution. Further, as junior enlisted personnel are generally not eligible for family housing on the base, they live off post, which draws much of their lives away from the military installation. Thus, there is a great potential for conflict between the greedy military institution's demands on junior enlisted personnel and the generally greedy demands of new marriages, which require intense interpersonal and psychological work.

Military Women

The number of military women has varied greatly, but the percentage has always been small. The largest number and concentration of women in the U.S. military, until recently, occurred in 1945, when approximately 265,000 women made up 2.2 percent of the force of over 12 million. During the 1950s and 1960s, women constituted less than 2 percent of uniformed personnel. Over the past 15 years, the number and percentage of women in the U.S. military have increased dramatically. At the end of fiscal year 1971, about 30,000 enlisted women and 13,000 female officers together constituted 1.6 percent of the total active duty military. By 1986, about 185,000 enlisted women and 30,000 women officers made a total of 215,000 in uniform (10 percent of the total active force).

Although women are still a small minority among military personnel, their presence in previously all-male military situations can potentially change the nature of social and interpersonal dynamics in those situations and may require institutional adaptation. Most relevant to analysis of military families is that the family has been normatively greedier for women than for men. For example, women have been expected to move where their husbands' jobs take them. However, military women, like military men, have less control than most civilians over where they will live, and they are unable to conform to traditional family expectations. The institutional adaptations here may come from the family as well as the military, with the family becoming less greedy for women.

Military women are less likely than military men to be married and to have children. In the navy, for example, active-duty women with civilian husbands constitute about 1.5 percent of total navy forces, and 69 percent of these women have no children. By contrast, among navy men with civilian wives, only 30 percent have no children. Married military women are also more likely than married military men to be separated from their spouses due to their military assignments. In the air force, for example, among military women married to civilian men, 10 percent of the officers and 36 percent of the enlisted women were geographically separated from their husbands. Although these couples constitute only 1 percent of all the armed forces, they account for 35 percent of married women officers and 23 percent of married enlisted women. This rate of separation caused by military assignments is much higher than for married military men with civilian wives.

Dual-Service Couples

A small but growing proportion of military families are in the special category of dual-service couples. These are married couples in which both husband and wife are in the military. Data from the 1978–79 DOD survey show that 3.9 percent of married enlisted personnel and 9.9 percent of married officers had spouses in the armed forces. The largest proportion was found among married air force officers, 13.1 percent of whom had military spouses.

As of 1984, there were 15,297 married army couples, which included army members married to uniformed personnel of other services. The percentages of army personnel married to other active-duty service members were reported by rank: 13.7 percent of first-term enlisted personnel, 6.5 percent of career enlisted personnel, 10.5 percent of company-grade officers, and 4.2 percent of field-grade officers.

Orthner and Nelson report that "some 2.5 percent of Navy personnel are married to other military members. In nine out of ten cases, the spouse is a member of the Navy. Most common are marriages between enlisted Navy women and enlisted Navy men." They also note that about "47 percent of all married Navy women are in dual-military marriages compared with only two percent of Navy men." They find that fully 70 percent of dual-military couples have no children.[21]

In the air force, dual-service marriages as of 1978 constituted 6 percent of the total force. In only 3.2 percent of these couples was the spouse a member of another armed force. The vast majority of dual-service marriages were between an enlisted man and an enlisted woman. Among married air force personnel, 76 percent of the women and 4.5 percent of the men had a military spouse. Among all dual-service air force couples, 69.4 percent had no children.[22]

Dual-service couples are often seen as a problem for personnel management because most such couples desire joint domicile, which requires the military to coordinate assignments of two people. The deployability of couples with children also causes concern because they, like single parents, do not have a civilian spouse expected to assume parental responsibilities.

Separation of dual-service couples due to assignments in different locations are more common than for other married military personnel. In the army in 1984, 38 percent of dual-service couples were not approved for joint domicile. In the air force in 1978, 12.5 percent of all dual-career couples were separated because of different assignments, ranging from 9 percent for spouses who were both enlisted personnel in the air force to 42 percent of air force women married to men in another service. Interestingly, "when separations by assignment occurred in dual-career military marriages, the children, if any, were as likely to stay with their fathers as with their mothers."[23]

Compared with couples in which one member is a civilian with career aspirations, dual-service couples may actually benefit the military organization in two major ways. First, although personnel assignment systems must be adapted for them, this may be accomplished more easily than coordinating with civilian employment opportunities. Second, and quite germane to the analysis of family effects on institutional and occupational trends in the military, couples consisting of two uniformed members are likely to be more committed to the military way of life and to understand each other's job requirements. As long as the armed forces can accommodate these couples, the military gains two members who are integrated into the organization, potentially reducing rather than increasing military-family conflicts. Orthner finds that "couples in which both the husband and wife are Air Force members are more likely to be happily married than couples with a civilian wife."[24]

Sole Parents

Sole parents (or single parents) are people who are currently unmarried but have custody of at least one minor child. Most sole parents are divorced or separated, although some have never been married. The proportion of military women who are sole parents is greater than the proportion of military men in that category. However, because military personnel are overwhelmingly male, there are more single fathers than mothers in the armed forces. It should be noted in analyzing trends in parenting among military women that, until 1975, military women (regardless of marital status) who became pregnant were automatically discharged from service.

Doering and Hutzler find that

> some 4 or 5 percent of the enlisted males in each Service have dependents but no spouse; among enlisted females, the percentage ranges from about 4 percent in the Air Force to 10 percent in the Army. Among officers, both males and females, the rates for dependents but no spouse never exceed 4 percent.[25]

Orthner and Nelson report that "almost one percent of Navy families are headed by single parents, approximately 4,500 members," that "2 out of 3 single parents are men," and that "nearly four percent of Navy women are single parents compared to less than one percent of the men." For the air force, "single parents currently comprise over 1% of the total force in the Air Force. Three out of four of these single parents are men."[26]

The family is greedy for sole parents because no other parent is available to share in childcare responsibilities. Conflict between military service obligations and family demands can therefore occur for these parents. Surprisingly, research specifically addressing this issue tends not to find greater difficulties among sole parents, at least as reported by the service members themselves. In his study of air force families, Orthner reports:

> Some military leaders have expressed alarm at the rising number of single parents in the services. Among Air Force single parents, concerns about their family stability, work performance, child care arrangements, and Air Force commitments appear to be largely unfounded. Some single parents do have problems, but compared to married couples, Air Force single parents are no more likely to report personal or parental stress, nor are they more likely to report work problems or weak Air Force career commitments.[27]

Changes in American Women's Roles

Increases in numbers of military women, dual-service couples, and sole parents are important changes in military family patterns. However, these patterns still characterize only a small minority of military families; the majority consist of a military man married to a civilian woman. Therefore, societal changes that affect civilian wives are likely to have the greatest impact on the military. Norms regarding the roles of American women, including wives, have been changing rapidly. Wives' personal and social identities are becoming increasingly independent of their husband's, and women's labor force participation rates have risen. These changes have been affecting military wives and will continue to force changes in military family relationships. Military wives are becoming less willing to adapt to the demands of the greedy military institution.

Data on labor force participation of wives of military men show them catching up to the rates for wives of civilian men but also show them having greater difficulty finding employment. Data from the 1978–79 DOD survey show 43 percent of enlisted spouses and 55 percent of officer spouses were in the labor force (working full- or part-time, or unemployed). U.S. Department of Labor data on military wives show that from 1970 to 1983, the percentage of working military wives increased from 31 to 48 percent. However, the unemployment rate is much higher for military wives than civilian wives: in 1983, 17.2 percent of military wives seeking jobs were unemployed, compared with 6.8 percent of civilian wives. In March 1984, military wives had a jobless rate of 17 percent, more than three times the national rate of 5.3 percent.[28]

Let us now consider actual and potential military adaptations to changes in military families, paying particular attention to their implications for institutional and occupational trends in the military.

MILITARY ADAPTATIONS TO MILITARY FAMILIES

Prototypical greedy institutions ensure the undivided loyalty of their members by normatively restricting outside social relationships from developing. The American military in the past limited the family ties of some of its personnel and continues to do so in various ways. Most notable is the exclusion from the service academies of anyone who is married or a parent. Control over the daily lives of enlisted personnel in basic training or boot camp restricts their contact with family members. Unaccompanied tours, although not prohibiting family ties, discourage personnel from bringing their families. During certain times of conscription, such as the early Vietnam years, drafting only young men and allowing exemptions based on family responsibilities limited the number of personnel with spouses and children.

Despite these mechanisms, the proportion of married military personnel has increased. This trend began even before the all-volunteer force was instituted in 1973 and has grown since then; given both the declining size of youth cohorts and public opinion regarding draft exemptions, even a return to conscription is unlikely to reverse it. Moreover, having more soldiers with families does not necessarily harm the military. They can be beneficial; for example, married soldiers have fewer disciplinary problems than single ones. Thus, the military institution has to adapt to the family patterns of its personnel. Such adaptation in the traditional military does not necessarily result in shifts toward occupational forms of service life; rather, some adaptations can preserve, or even heighten, institutional commitment.

Recruitment, morale, and retention of military personnel are affected by family members' attitudes toward the military life-style. Thus, the armed forces have been concerned with the quality of family life and must continue to address changing family patterns. In response to both the recognition of family problems and direct pressures from military family organizations and advocates, the services have been developing family programs and considering potential impacts on families in making personnel policy decisions. The directions of these adaptations can be analyzed for their effects on the institutional nature of military organization.

The military began providing a variety of family support services in the 1960s, and these programs have been expanded in recent years. Organizations such as Army Community Services, the Navy Personal Services Centers, the Air Force Family Support Centers, and other organizational subunits offer assistance directly to families, which includes legal advice, family counseling, financial help, English classes for foreign-born spouses, and relocation assistance. Relocation services include advance information about the area and installation to which a service member is being transferred, assigned sponsors at the new location, and temporary loans of household items. Family advocacy programs have been established to prevent and treat cases of family violence (child and spouse abuse).

Among the organizations and activities affecting military family policies is the National Military Family Association (originally named the National Military Wives Association), which acts as an advocate for military families in the services and in Congress. The Army Family Committee, with the support of the Army Officers Wives Club of the Greater Washington Area and the Association of the United States Army, and later the army itself, organized several Army Family Symposia in Washington, D.C., in the early 1980s. These meetings brought together family representatives from army installations all over the world to discuss problems, share information on effective family support programs, and make recommendations to the army.

Responses of the military to family concerns and pressures include establishment of a family adviser to the assistant secretary of defense for manpower, installations, and logistics. The Army Office of the Deputy Chief of Staff for Personnel has a Family Liaison Office to represent family concerns. The army chief of staff declared 1984 the year of the family. Although the programs developed have not yet created major institutional changes in the demands the military makes on service members and their families, they have established mechanisms for identifying and addressing areas where change is desirable and possible, and voices within the military organization represent family concerns.[29]

Military adaptation can potentially increase the institutional commitment of service members and their families in several major areas of concern considered in this chapter. These include junior enlisted families, spouse employment, military women, and dual-service couples.

The special circumstances of young enlisted families place them at high risk for a host of problems, including poverty, family violence, psychological adjustment difficulties, and marital dissolution. Appropriate policies can ease their burdens while at the same time integrating the families into the military institution. For example, provision of on-post housing can relieve some of their financial pressures, facilitate their use of available on-post services, and socialize them to the military way of life. Further, even without providing on-post housing, the armed forces can offer opportunities for junior enlisted families to develop informal social support networks that ameliorate the effects of stressful life events. Personnel policies such as the army's unit rotation can facilitate these opportunities if the units make specific attempts to do so. Incorporating enlisted wives into the military organization in ways similar to the institutionalization of the roles of officers' wives can potentially strengthen their husbands' commitment to the military. Indeed, the direction of social change with regard to women's roles, and the effects of social class on these changes, may presage more integration into post life for enlisted wives and less for officers' wives.

The services have begun to accommodate the employment needs of military spouses, but much remains to be done. Among the policies that can be beneficial are longer tours in one location, job banks, education and job training services, expanded childcare facilities, and coordination with civilian employers (including those overseas near military installations). Further, the armed forces need to actively oppose local hiring policies that discriminate against military spouses, even for government jobs in the military community, because these policies seriously impede employment.

The increases in women in uniform have already required changes in the traditionally male military institution. Such organizational adaptations, however, do not make the military any less of an institution. Military women generally enter the armed forces for the same reasons as men. Indeed, they may be even more institutionally committed than men because women, socialized to subjugate their individual goals to group goals, in general are often found to be more service- and collectively-oriented. However, the morale and career commitments of military women are affected by their treatment within the organization. Discussing all the ways to improve the integration of women in the military is beyond the scope of this chapter. However, specifically regarding family issues, given that the family is normatively greedier for women, service policies can aim at being partially accommodating to especially greedy family stages. For example, military members can be allowed to take parental leave, and childcare programs can be improved.

The potentially high commitment of dual-service couples has already been noted. To retain these couples, the major effort required by the military is the provision of coordinated assignments. Although doing so gets more difficult as the numbers of such couples increase, the gains in institutional dedication may be worth the accommodations necessary.

In general, the more the military service adapts to family needs, the more committed will be both service members and their families to the institution. Family quality of life and satisfaction with the military can be improved by such policies as, when possible, taking family desires into account in the timing of relocation, giving advance notice of transfers, and providing various forms of assistance when separations are necessary.

The more the military's actions make service members and their families truly hear and believe the message that "the military takes care of its own," the less will be the conflict between the two greedy institutions of the military and the family. To the extent that the military views the family as an outside influence with which it competes, the military will likely move in an occupational direction. To the extent that the military works to incorporate the family within itself and adapts to it, the result will be institutional change but preservation of the institutional nature of military organization.

NOTES

1. Lewis A. Coser, *Greedy Institutions: Patterns of Undivided Commitment* (New York: Free Press, 1974).
2. Ibid., p. 3.
3. Ibid., p. 6.
4. Lewis A. Coser and Rose Laub Coser, "The Housewife and Her 'Greedy Family,' " in Lewis A. Coser, ed., *Greedy Institutions: Patterns of Undivided Commitment* (New York: Free Press, 1974), pp. 89–100.
5. Ibid., p. 100.
6. For reviews of research and annotated bibliographies on military families, see Willard M. Bennett et al., *Army Families* (Carlisle Barracks, Penn.: U.S. Army War College, 1974); Susan Farish, Francoise Baker, and Marilyn Robertson, "Research on the Military Family: An Annotated Bibliography," in Hamilton I. McCubbin, Barbara B. Dahl, and Edna J. Hunter, eds., *Families in the Military System* (Beverly Hills, Calif.: Sage, 1976); Edna J. Hunter, *Families Under the Flag: A Review of Military Family Literature* (New York: Praeger, 1982); Edna J. Hunter, Donald Den Dulk, and John W. Williams, *The Literature on Military Families, 1980: An Annotated Bibliography* (Colorado Springs, Colo.: U.S. Air Force Academy, TR No. 80-11, 1980); Military Family Resource Center, *Review of Military Family Research and Literature*, vol. 1: *Review* and vol. 2: *Annotated Bibliography* (Springfield, Va.: Military Family Resource Center, 1984).
7. Zahava D. Doering and William P. Hutzler, *Description of Officers and Enlisted Personnel in the U.S. Armed Forces: Reference for Military Manpower Analysis* (Santa Monica, Calif.: RAND, 1982), p. 161. The 1978–79 survey contains the most recent comprehensive data publicly available. Although rates of occurrence of some organizational demands may have changed somewhat since the survey was conducted, the changes are unlikely to be large enough to affect any of the conclusions drawn here. Indeed, these data can serve as a baseline to map changes in the demands on service members and their families.
8. Ibid., p. 168–169.
9. See, for example, Jerry L. McKain, "Alienation: A Function of Geographical Mobility Among Families," in McGubbin, Dahl, and Hunter, eds., *Families in the Military System*, pp. 69–91; Mady Wechsler Segal, "Enlisted Family Life in the U.S. Army: A Portrait of an Army Community," in David R. Segal and H. Wallace Sinaikio, eds., *Life in the Rank and File: Enlisted Men and Women in the Armed Forces of the United States, Australia, Canada, and the United Kingdom* (McLean, Va. Pergamon-Brassey's, 1986), pp. 184–211; John D. Woelfel and Joel M. Savell, "Marital Satisfaction, Job Satisfaction, and Retention in the Army," in Edna J. Hunter and D. Stephen Nice, eds., *Military Families: Adaptation to Change* (New York: Praeger, 1978), pp. 17–31.
10. Paul Darnauer, "The Adolescent Experience in Career Army Families," in McCubbin, Dahl, and Hunter, eds., *Families in the Military System*, pp. 42–66.
11. Department of the Army, *Soldiers Report*, pp. 8–7; Allyson S. Grossman, *The Employment Situation for Military Wives* (Washington, DC: U.S. Department of Labor, Bureau of Labor Statistics, 1984).
12. Reuben Hill, "Foreword," in McCubbin, Dahl, and Hunter, *Families in the Military System*, p. 13.
13. Doering and Hutzler, *Officers and Enlisted Personnel*, pp. 130–31.
14. For a review of some of this literature, see Hunter, *Families Under the Flag*, pp. 44–50; and the

literature cited in Mady Wechsler Segal, "The Military and the Family as Greedy Institutions, *Armed Forces and Society* 13 (Fall 1986): 9–38.

15. Doering and Hutzler, *Officers and Enlisted Personnel,* pp. 180–184.
16. Charles C. Moskos, "Institutional and Occupational Trends in the Armed Forces: An Update," *Armed Forces & Society* 12 (Spring 1986): 381.
17. Nancy Shea (and revised by Anna Perle Smith), *The Army Wife* (New York: Harper & Row, 1941 to 1966).
18. Ibid., p. 1.
19. Ibid., p. 55.
20. Nancy L. Goldman, "Trends in Family Patterns of U.S. Military Personnel During the 20th Century," in Nancy L. Goldman and David R. Segal, eds., *The Social Psychology of Military Service,* (Beverley Hills, Calif.: Sage, 1976), pp. 119–120.
21. Dennis K. Orthner and Rosemary S. Nelson, "A Demographic Profile of U.S. Navy Personnel and Families." Prepared for the Navy Family Support Program, August, 1980.
22. Richard J. Brown III, Richard Carr, and Dennis K. Orthner, "Family Life Patterns in the Air Force," in Franklin D. Margiotta, James Brown, and Michael J. Collins, eds., (Boulder, Colo.: Westview, 1983), pp. 207–220; see also Dennis K. Orthner, *Families in Blue: A Study of Married and Single Parent Families in the U.S. Air Force* (Washington, D.C.: Office of the Chief of Chaplains, U.S. Air Force, 1980).
23. Ibid., pp. 213–214.
24. Ibid., p. 214.
25. Doering and Hutzler, *Officer and Enlisted Personnel,* p. 95.
26. Orthner and Nelson, "Demographic Profile of U.S. Navy Families: Orthner, *Families in Blue.*
27. Ibid., p. 113.
28. All data in the paragraph from Bureau of Labor Statistics (BLS), *Monthly Labor Review,* February 1981, and updated statistics from BLS via personal communication with MS. Waldman.
29. For a history and description of these programs, see Gary Lee Bowen, "Military Family Advocacy: A Status Report," *Armed Forces & Society* 10 (Summer 1984): 583–596.

VII

Sex Roles in the Military

PATRICIA M. SHIELDS

A striking characteristic of the modern U.S. military is its increased reliance on women, an unplanned consequence of the draft's demise. Currently, American women serve their country in unprecedented ways and numbers. No other industrialized nation has ever used women as extensively. As peacetime soldiers, they are unlike their historical sisters who were called to meet the challenge of war.

Overall, women make up approximately 10 percent of the active force, representing more than a fivefold increase since 1970. Not only is this growth unmatched in history, women are also assigned to an increased variety of occupational specialties, including combat support. According to reports, women also tend to raise the quality of the armed forces; enlisted women, on average, have more education and score higher on standardized tests.[1]

Some view the new widespread use of women as a better and more equitable military personnel policy.[2] Others view the trend with alarm. Skeptics, who see the increased proportion of women as a radical step, fear that readiness and combat effectiveness will be sacrificed in favor of an ill-conceived social experiment.[3]

Wherever the truth lies, female participation is unlikely to return to prevolunteer force levels (under 2 percent). Indeed, official U.S. armed forces projections suggest greater reliance on women.[4] Hence, an understanding of the appropriate use of women is critical in arriving at an effective military human resource policy. What kinds of values and norms do women bring to the military? Are they economically motivated, joining the

Many thanks to Tim Carr, Daniel Hesser, Capt. Mary Gueurts, and Bill Childers for their comments on an earlier draft. I would also like to thank M. B. Freels for her research assistance.

military to maximize their long-run earnings stream, or are they like their dedicated, loyal counterparts of the 1940s and 1950s?

Another important set of questions addresses the effect of their increasing numbers in the military. Will swelling numbers of women alter traditional military norms and values? Does the institution help a woman feel a part of a team with an important mission, or is she shut out, alienated by an organization steeped in male tradition? Clearly, the answers to these questions are evolving. However, two issues emerge: one concerns the values of women as they enter and remain in the military; the other focuses on the organization and how it shapes these values.

This chapter will address the role of military women in light of these issues, and two findings become clear. First, women do not consider the military as just another employer; rather, they are attracted to unique aspects of the military institution, such as discipline and adventure. Second, women soldiers are not fully supported in their institutional identities. Institutional attachment—often eroded by family responsibilities—is diluted because the military does not really accommodate women.

Women's presence has been expanding in both the military and the work force. For example, in 1962 only 37 percent were members of the labor force; more than half were full-time homemakers. By 1982 these proportions had reversed: 35 percent were keeping house and 53 percent were employed or looking for work. Further, between 1972 and 1982, as female labor force participation grew, there was a dramatic jump (57 percent) in households headed by women. One in six families is now maintained by women.[5]

Historically, women's association with the labor force has been continuous but limited. In the civilian sector, women have never been totally excluded from paid employment. Female military participation, however, has been more sporadic. During wartime, women have enlisted, performed traditional female tasks, and freed men for combat. At war's end, all but a few invariably returned to civilian life.

Because revolution, invasion, and violent world conflict demand that citizens work together and sacrifice to ensure survival, reluctant military institutions have sometimes called on women to add their talents to war efforts. However, women's participation in the military system of industrial societies has been narrowly limited to traditional fields and activities, where their presence did not threaten men or male traditions.

For example, women were first allowed to contribute by nursing the wounded (e.g., England during the Crimean War). Later, during World War I and World War II, women freed men to fight by taking on other traditional jobs such as typing, sorting mail, and laundry. (We should note that during war, these traditional tasks differed from peacetime civilian equivalents.) Women recognized that their work contributed to the war effort, and they were committed to something much larger than just a job. Military duty was not easy; however, despite long hours, dangerous and sometimes filthy working conditions, poorly designed and ill-fitting uniforms, smear campaigns, living quarters that approached house arrest (in the South Pacific), and patronizing and harassing men, women served their country with dedication.[6]

After World War II, American women were excluded from service directly by quotas and indirectly through policies that mandated their discharge upon marriage or pregnancy. These policies resulted in first-term attrition rates sometimes as high as 80 percent during the 1950s.[7] The small cadre of women who remained developed a strong attachment to the institution.

Helen Rogan interviewed and observed former members of the Women's Army Corps (WAC). Her account suggests that these women were proud of the WAC. They described an element of "super giant, good military bearing with everything right. It is when you are dressed . . . , polished, and spit-shined at all times."[8]

Rogan maintained that women of the army corps had an allegiance to an institution within an institution. The small, isolated WAC world was unlike the world of work. It became part of a woman's personality in a way that no civilian job could. "The sense of mission suffuses her work. Women of the Army Corps had little in the way of status or recognition to sustain them, [and] this tended to make their sense of mission even stronger."[9]

After World War II, the assumptions surrounding national security changed. The highly technical nuclear era led to new national security goals such as containment and deterrence. This dictated a shift to immediate readiness as an armed forces policy goal. Immediate readiness necessitated a large standing force, a mighty war machine whose effectiveness is judged by its ability to maintain peace through power. During the 1950s and 1960s, the armed forces relied upon the draft to meet its large manpower needs.

The young all-volunteer force, experiencing a manpower shortfall, began to rely increasingly on womanpower. At the same time, the 2 percent ceiling on female participation was lifted. Armed forces personnel demands opened the doors for female integration. Early on, the separate women's services were dismantled. Although fast paced, integration had many false starts and setbacks. By the mid-1980s, only a small percentage of military jobs were closed to women due to combat restrictions. Col. Evelyn Foote, a WAC officer who lived through the changes, described them as "evolutionary changes in a revolutionary time frame."[10]

INSTITUTIONAL OR OCCUPATIONAL ORIENTATIONS OF WOMEN

When the military dropped conscription and allowed individuals to freely choose the military, it also adopted a labor market recruitment model. Under this model, the military became an employer competing for human resources in the marketplace; the soldier became a rational employee who chooses the military *occupation* to maximize long-term earnings. Wages, in this context, are assumed to be the critical enlistment motivator.

Clearly, the labor market model contrasts sharply with traditional military recruitment systems. Traditional methods used institutional values—duty, honor, country, and esprit de corps—to motivate enlistment and maintain quality recruits. The draft, the cornerstone of the institutional model, met staffing demands directly and motivated enlistment as well. In addition, military service was viewed as an obligation. Public support for the draft rests upon citizen acceptance of service as a national obligation.

When the Vietnam War eroded public support for the draft, it ushered in the use of labor supply models. Paradoxically, when the military relied upon the draft, it held its most institutionally oriented segment—women—at arm's length. The military employer, on the other hand, opened doors to women and used them in ways unparalleled in history.

The traditional and labor market frameworks (discussed previously) have been described as the institutional-occupational dichotomy by Charles Moskos. Although little survey evidence distinguishes men from women along I/O dimensions, women have historically been very institutional. Obviously, the women soldiers of World War II met WAC leader Oveta Culp Hobby's challenge of a "debt to democracy" and a "date with destiny."[11] Further, in their separate and tiny women's service corps, the career women of the 1950s and 1960s are assumed to have had a strong institutional attachment. (How many male career service personnel would there be if marriage resulted in dismissal?)

The notion of self-sacrifice is not alien to women. In their role as mothers, women understand this concept all too well. In addition, traditional occupations such as nursing and teaching have been viewed as callings. The former is a prime example of a traditional female vocation compatible with military institutional values. Indeed, the concept of nursing as a calling has been used and acknowledged by the armed forces. When describing American nurses during the Spanish-American War, Nicholas Senn depicted nursing as a woman's "special calling. . . . Her sense of duty of devotion was seldom matched by men."[12] Recruitment posters during World War II gave military nurses nunlike qualities; captions termed them "Greatest Mothers in the World."[13]

Historical evidence that female soldiers treat military service as a vocation is strong. These women, however, represented only a fraction of available American women. Indeed, during World War II, the military was unable to meet its female recruitment goals. Nurse shortages were so severe that President Roosevelt called for and the House passed a nurses' draft.[14] Further, the high attrition rates during the 1950s and 1960s indicate that 70 percent to 80 percent of the women initially attracted to the military did not find the feeling of obligation or mission strong enough to complete their first tour of duty. Hence, although historical evidence supports the notion of a calling, it also suggests that the military cannot rely on this factor alone to attract and keep the number of women it requires in the late 1980s and the 1990s.

What of the modern woman soldier in the new integrated military? Does she consider the armed forces a vocation or occupation? At first glance, the average woman recruit would appear to be motivated by occupational norms. From an employment perspective, little practical distinction should exist between the military and civilian sectors because, for women, the military and civilian occupational structures closely mirror one another.[15] This alone, however, does not imply that women are attracted to the armed forces for occupational reasons. If women are occupational, they will enlist out of self-interest to maximize their net economic advantage. In other words, they will enlist to take advantage of relatively high wages, fringe benefits, and job security.

According to the occupational thesis, variables such as wages and unemployment are critical in explaining enlistment, attrition, and reenlistment. If these variables are effective at predicting female military participation patterns, they would certainly support the contention that women are attracted to the armed forces for occupational reasons.

The military's wage package is very attractive to women because in the civilian workplace women tend to be clustered in low-paying occupations. Further, because job security in the military is greater than in civilian life, a woman can expect higher long-term earnings in the armed forces.[16] The military also offers an attractive training package. As the country's largest training institution, the armed forces provide both on-the-job and formal instruction to enlistees. Curricula include technical fields with significant civilian transferability, such as electronics.[17] Unemployment is another variable to be considered. Obviously, periods of unemployment erode the net economic advantage of civilian employment. When the U.S. economy is losing ground, the military is a source of steady employment. Not surprisingly, during periods of high unemployment, the AVF has found recruiting quality soldiers easier. Historically, women's unemployment rates have been higher than men's.[18] Hence, the entire package of relatively higher wages, job security, and training should make the military an attractive employer to women.

Enlistment

Women's enlistment rates cannot be used to assess their occupational orientation because it is influenced by military demand. It does not incorporate the number who want to enlist but are turned away. The desire to enlist or enlistment intention is a more accurate measure of occupational orientation.

Given the clear net economic advantage associated with military employment, the occupational model would predict enlistment intention rates equal to or greater than overall labor force participation. The evidence, however, fails to support the occupational model. Women expect to enlist in percentages that are a small fraction of female labor force participation. Between 1976 and 1982, the proportion of female high school seniors who expected to enlist ranged from a high of 7 percent (1976 and 1982) to a low of 5.4 percent (1979 and 1980).[19]

What of the women who do enlist? Are they motivated by occupational factors? In a study of a large national sample of enlisted men and women, "pay" and "unemployment" were tied for last (among 12 choices) as the reason for female enlistment.[20] Accordingly, these women rejected occupational values as enlistment motivators. Further, these women cited "to better myself in life" as the major enlistment motivator. What do young women mean when they check "to better myself in life"? Obviously, they were not referring to training or educational benefits because both were included as choices in the questionnaire. The discipline and structure of the military institution both seem of special interest to young women.[21]

Recent field interviews also suggest that pay is not the primary reason women enlist. Moskos found that army women on field duty in Honduras gave reasons for enlisting such as "sense of adventure" or "get away from boring community."[22] Men, on the other hand, listed economic incentives. In another exploratory study among army, marine, and air force women, Shields discovered that military women do not like to classify or rank their reasons for joining. Instead, they said they were attracted to the military because it offered new and exciting challenges. They enjoyed the thought of seeing the world and viewed themselves apart and different from civilian women. Theirs was *not* just another job—it offered excitement, adventure, discipline, and structure. They seemed to be attracted to the military because it had institutional qualities unlike civilian employment. Nevertheless, despite the chance for an exciting life, a large part of their enlistment decision rested with the need to find employment, support themselves, and enter the adult world.[23]

Reenlistment Decisions and Attrition

Like enlistment decisions, reenlistment decisions and attrition rates of women are important indications of their institutional or occupational orientation. Compared with men, women are more likely to leave the military before the end of their first tour. Approximately 45 percent of the enlisted women who entered in 1981 did not complete their first tour. This compares with 34 percent for men.[24] Perhaps this is not surprising because women have traditionally had higher labor turnover rates than men.[25] However, military women's turnover rates are lower than those of their civilian counterparts, who are twice as likely to leave their employers in a given year.

Virtually the entire difference in attrition between military men and women is attributable to women's increased likelihood of leaving prematurely for pregnancy and parental responsibilties.[27] For example, among women who entered in 1980, 13 percent left for family responsibilities; only 2 percent of the men, on the other hand, felt fatherhood was incompatible with enlisted life. Significant numbers of enlisted women appear to find mixing children and military life (be it vocation or occupation) difficult.[28]

Women of the army and marines have the highest attrition rates, perhaps because these services are most closely associated with hand-to-hand combat and other disamenities not usually found in the labor market. Female attrition is the lowest in the navy, approximately that of men.[27]

The situation is different for women who complete their first tour of duty. Women's first-term reenlistment rates have consistently run somewhat higher than men's and have been increasing since 1981. However, career women reenlist at rates substantially lower than career men. For example, in 1983, approximately 90 percent of the career enlisted men reenlisted, compared with only 72 percent of their female counterparts.[30]

The evidence that economic incentives motivate women to complete their first term or to reenlist is mixed. Kim found that job satisfaction was the only significant variable positively associated with first-term reenlistment intentions. Good pay was ranked as the most important component in job satisfaction, followed by learning valuable skills and pleasant physical surroundings, ranked second and third, respectively.[31] The larger role of military pay is perhaps not surprising. Reenlistment implies that the military is considered a career, not just a place to "get away from a boring community" or to learn about discipline and structure. These goals would have been fulfilled in the first tour. The importance of pay overall, however, should not be overstressed.

In a study that looked at the determinants of first-term reenlistment intentions, job satisfaction was found to be statistically significant, while pay itself was not.[32] Hence, pay appears to influence reenlistment intentions indirectly through job satisfaction. Thus, although not a primary factor in explaining commitment to the institution, pay is more important later than at initial enlistment. In addition, women planning to leave do not cite poor economic conditions as their primary source of dissatisfaction. "Unpleasant physical surroundings" is ranked number one.[33]

Reenlistment decisions provide further evidence that women find it difficult to balance family and military responsibilities. Controlling for other factors such as job satisfaction, marriage, and traditional values, Kim found that the presence of a child decreases reenlistment probability by 69 percent.[34] As one might expect, enlisted mothers with traditional attitudes are not interested in combining a military career and children. Children are enough of a stressor that, for traditional and nontraditional women alike, on average, their reenlistment propensities decline.

Most of the studies discussed above dealt with enlisted women. Relevant literature on officers is more limited. One would expect officers to hold institutional values because they are members of the military's leadership. For example, the much-studied female academy students evidence strong institutional attachment. In a recent survey of Fort Hood women officers, pay proved to be a relatively unimportant enlistment motivator. Although the respondents felt military compensation was adequate, a majority believed that their long-and short-run earnings would be higher if they were civilians. Other factors were more important, for example, they trusted their leader and were satisfied with the challenging work. Moreover, the overwhelming majority favored regulations regarding dress and appearance, valued discipline, and felt they were members of a prestigious organization.[35]

Institutional orientations tend to be strongest among career military. Although women are currently underrepresented in the higher ranks, they —like their male peers—will probably become more institutional as they move up. Another factor that may affect women's orientation is their specialty. Because women are excluded from combat specialties, where even among the career force institutional values are highest, one would expect future institutional values to be weaker among women even if they are relatively senior.

Results of enlistment, reenlistment, and attrition studies suggest that female soldiers are not concerned exclusively with net economic advantage. Pay is important, but pay alone does not determine whether a woman enters or remains. Trends in female attrition and reenlistment suggest that the military is weeding out those who are dissatisfied, those who fail to meet minimal criteria, and those who are unwilling or unable to balance both family and military responsibilities.

The evidence presented here, although not conclusive, leads one to believe that military women, in general, do not enter or remain primarily for occupational reasons. They are attracted to an institution, not an employer. They tend to view the military as a special institution offering unique benefits, such as discipline and adventure. The behavior and attitudes of men, on the other hand, are more consistent with the occupational theses. However, a small but significant number of men do hold strong institutional values. They are primarily career soldiers in combat-related specialties. Hence, with the exception of males in combat roles, female enlistees probably hold values more compatible with the institutional mode.

THE EXPERIENCE OF WOMEN IN THE MILITARY

The armed forces have their roots in time-honored masculine traditions. The uniforms, rituals, and authoritarian structure that permeate military life help transform boys into effective men soldiers. Many traditions and practices of the male-oriented institution clash with effective female assimilation. —

Taken as a whole, men of the armed forces have resisted and been hostile toward attempts at gender integration. The need to use women is often recognized first at the top of the command structure. Historically, however, initial attempts at female integration have not been accepted at the day-to-day level. Generally, the rank and file have resisted and obstructed integration. During World War II, a spontaneous campaign arose among American GIs to sabotage the new women's military corps, and military men engaged in a slander campaign against their female counterparts, targeting morals and character.[36] Hence, initial integration was resisted by men, even when women were assigned to traditional tasks such as nursing, typing, and filing.

Scholars speculate that men resist initial integration because they find it threatening. The very fact of a woman succeeding in the military dilutes time-honored male rituals and reflects poorly upon the concept of the dominant male. Essentially, they have invaded an elite (perhaps mystical) male testing ground.[37]

Among the potentially most debilitating obstacles military women face is sexual harassment. In an extensive study of Signal Corps women in Khaki Town, West Germany (comprising four American army bases), Michael Rustad describes its devastating effects. He found sexual harassment to be pervasive, constant, and demoralizing.[38] Obviously, people treated as sexual objects are seldom accepted as part of a cohesive team. Sexual harassment erodes institutional loyalty and undermines values.

In the modern, volunteer military, the most severe problems of integration occur when women work in nontraditional specialties. Unlike their counterparts in pink-collar jobs, these women experience all the problems common to tokens. The male enlisted culture is a working-class machismo—in Rustad's Khaki Town, the adventure comic book was the number-one best-seller—women in nontraditional assignments threaten this culture.

Women in nontraditional MOSs (military occupational specialties) often find themselves in a double bind: when they succeed in their work role, their femininity is questioned; when they fail, their womanliness is affirmed at the expense of their work role. The problem is described very well by a maintenance specialist in Rustad's Khaki Town.

> To me, men really can't handle women in jobs like these. They try and instill an attitude in the females that they don't know what they're doing. Once this is done, they come in and offer help. Then they say women can't do the job.[39]

Given the hassles associated with nontraditional jobs, it is not surprising that women in these slots, on average, score lower on scales of job satisfaction and are more likely to leave the institution.[40] In the first place, enlisted women generally are not attracted to nontraditional jobs. Typically, they value a cheerful, cleanwork environment; they are less interested than men in working outdoors, with their hands, or with machines. In reality, the typical young woman is less prepared than the average man, both psychologically and physically, for any of the traditional male occupational specialties.[41]

Today, a number of factors have helped to create a more supportive environment in the military and to reduce tension between the sexes. They include time, increased numbers of women, better matching of physical capabilities with jobs, changing expectations of youth cohorts, a greater sense of patriotism, more egalitarian attitudes toward women's role in society, and explicit military programs and directives dealing with sexual harassment.

With fuller integration, people are more likely to judge one another as individuals. Day-to-day integration is working. It is "creating new levels of mutual trust and confidence."[42] Two recent studies examined female integration under rigorous and dangerous field conditions.[43] Field assignments have few civilian employment counterparts. Almost by definition, successful field activities rely on institutional values. Field success and survival depend, in large part, on unit cohesion, which is fostered through institutional values such as fellowship, trust, and esprit de corps.

The authors of both studies observed that the men and women worked well together. Over the course of field training, women were increasingly judged as individuals. Buddy or brother-sister relationships were the norm, not romantic attachments. Men and women built cohesion through shared experiences. The distinction between insiders (those within the unit) and outsiders was more important than that between the two sexes. Individuals were judged by how effectively they performed the task and were prized for their knowledge.[44]

Within units, there was little sexual harassment. Hence, under the very conditions in which sexual harassment could be most devastating to the military mission, it did not seem to be a problem. Nevertheless, men outside the immediate unit, particularly those with low-status jobs (e.g., cooks) and poor educational backgrounds, continued to harass women. Although sexual harassment seems to be less of a problem, it has not disappeared completely.

Nonetheless, many of the issues that alienated women and eroded institutional values in the 1970s have been addressed, and young women soldiers of the 1980s are more likely to have experiences consistent with institutional values. The big issues yet unresolved for military women in this decade seem to be pregnancy, child care, and combat exclusion, all of which affect the utilization of women in the military.

Pregnancy and child care are two of the most controversial and emotional issues associated with female integration. Pregnancy is a very troubling issue for women in nontraditional jobs, particularly combat support and combat service support, because it has implications for deployability. Pregnancy disrupts unit cohesion and team effectiveness in a uniquely feminine way. These issues are likely to be viewed differently by men and women in the unit. Pregnant women are apt to be resented by men because of the special burden they bring to the unit: men may be fathers, but they do not bring their wives' pregnant condition to work with them. They do not have poor balance, nor must they avoid heavy lifting or certain chemicals for several months. In short, whether or not men choose to have children does not affect their ability to perform tasks (unit cohesion) in the same way.

Focusing on lost time due to pregnancy is a false issue, according to Senator Proxmire. Men have lost more time for abusing drugs and alcohol than women have for drug and alcohol abuse and pregnancy combined.[45] From the deployability point of view, however, the man with a hangover and the pregnant women both present problems. Just as all men are not banned because some get hangovers, all women should not be banned because some get pregnant. Also, all pregnant women are not a burden in all (or most) jobs. Nevertheless, from a strict deployability standpoint, pregnancy represents a short-term burden in some military functions.

Parenting, a family function that is shared by men and women soldiers, has special implications for women. Motherhood takes more time and energy because women usually organize household activities and are responsible for a majority of childcare duties. (Both father and mother change diapers; mom, however, buys them.) Although large numbers of women choose to balance military careers and families, analysis of attrition and reenlistment trends show that motherhood is one of the major reasons women leave. In contrast, being a father increases a man's probability of reenlisting.[46] This evidence suggests difficulty for women in keeping up with the demands of military life and motherhood. Women, on average, are more likely to leave nontraditional career fields, which often have irregular hours. This suggests that the mixing of motherhood and a military career is most difficult when women work in an MOS that has unconventional hours and long trips away from home not compatible with accessible child care. When a women's specialty area has regular hours and weekends off (e.g., air force legal work) childcare needs would be similar to those of civilian counterparts. (In addition, such military mothers would have the support necessary to both advance in a military career and raise a family.)

Clearly, military women are young and in their childbearing years. They also tend to value children and family life. For example, Shields found, using a national sample, that over 95 percent of the enlisted women believed the ideal family contained two or more children. Moreover, 85 percent *desired* two or more children.[47] Motherhood is clearly a demanding, time-consuming activity. Hence, it is relevant to examine just how it might influence I/O orientations. One might speculate that mothers who are also soldiers would have a tendency to adopt occupational values. Such values would allow a woman to maintain the primary sense of duty to her family and treat the military as a job. On a day-to-day basis, satisfying both military duty and motherly obligation may be fairly easy. Small children, however, have suppressed immune systems and at day-care centers are likely to get more infectious diseases than children cared for at home or in small groups.[48] If forced to choose between military responsibilities and a sick toddler, most mothers will choose the child. Moreover, single parents must take the time to deal with these problems without the support of a spouse.

Hence, child care for dependents of active duty personnel is an important issue among current military personnel.[49] In a group interview held at Kelly Air Force Base (February 1985), an interesting, very institutional, perspective was taken by career single mothers. These women lived on base and used the reasonably priced base day-care facilities. They also worked in traditional specialties with regular hours and had weekends off. They liked the air force because it was an "institution that took care of its own." They were patriotic and enjoyed being members of the armed forces. They emphasized the benefits of free, accessible medical care, job security, convenient day care, and the knowledge that they are safe on base. They never feared rape or attack while at Kelly. Given the desperate plight of many civilian single mothers, that these military mothers adopted this perspective is not surprising.

Women are integrated into almost every aspect of military mission except direct combat activities and draft registration. In the combat arena, women will probably die in proportion to their numbers in a major ground conflict. They will not, however, be among the "bands of brothers" who will be called to do the initial killing. Not surprisingly, the institutional orientation is highest among career combat soldiers. The values that tie these men together and produce unit cohesion are institutional. Hence, as long as women are excluded from this inner circle, their institutional orientation on average, will, not reach its full potential.

In the United States, women have always been excluded from the draft. From an institutional perspective, the draft is viewed as an obligation to serve the nation. In the ideal sense, it forces a few (and motivates many) to serve their country. Today a male citizen of the United States has the obligation to register for the draft. Women do not. Neither the civilian leadership nor the military institution has asked women to view service as an obligation. Thus, the military is giving women a message that undermines female institutional values and, by implication, promotes an occupational orientation.

CONCLUSION

Women today are an integral part of the U.S. armed forces. They do not view the military as a civilian employer. Their heritage is steeped in institutional values and they are institutionally oriented. Economic factors are important to them, but they are also attracted to the structure, tradition, rituals, discipline, and opportunities of the military. Further, the typical female enlistee is probably more institutionally oriented than her male counterpart. Reenlistment is also motivated by institutional values.

Women have not been easily assimilated into the military organization. There are still serious problems of adaptation. High female first-term attrition

(45 percent) is symptomatic. Perhaps the problem rests with a military organization that does not really welcome or accommodate women. Sexual harassment and poorly designed apparel (e.g., ill-fitting boots) are but two examples. In addition, problems are most apt to occur when women work in nontraditional specialties.

Motherhood also disrupts assimiliation. This is particularly true when women work in occupations with irregular hours and extended absence from home. Women who want full careers in these fields will probably not be mothers. Motherhood is less problematic when women work in jobs with more regular hours.

The military has an opportunity to build institutional attachment among women (and all active-duty parents) by providing more family-oriented noncash benefits—such as expanded, flexible, quality child care and dependents' quarters. Using innovative institutional methods, the armed forces could nurture institutional attachment in the modern era of working mothers, but is that a choice they are likely to make?

NOTES

1. Martin Binkin and Shirley Bach, *Women and the Military* (Washington, D.C.: The Brookings Institution, 1977), p. 17.
2. See, for example, Nancy Goldman, ed., *Female Soldiers—Combatants or Noncombatants* (Westport, Conn.: Greenwood, 1982); Jeanne Holm, *Women in the Military: An Unfinished Revolution* (Novato, Calif.: Presidio Press, 1982); Helen Rogan, *Mixed Company: Women in the Modern Army* (New York: G.P.Putnam's Sons, 1981); Mady W. Segal, "Women in the Military: Research and Policy Issues," *Youth and Society* 10 (December 1984): 33.
3. Michael Levin, "Women As Soldiers—The Record So Far," *The Public Interest* 76 (Summer 1984): 33.
4. Department of Defense (hereafter DOD); Manpower, Installations, and Logistics, *Military Women in the Department of Defense*, II (Washington, D.C.; April, 1984), 3.
5. Department of Labor (hereafter DOL), *Women at Work: A Chartbook* (Washington, D.C.: Bureau of Labor Statistics, 1983).
6. Holm, *Women in the Military*, pp. 9–12.
7. Ibid., p. 163.
8. Rogan, *Mixed Company*, p. 14.
9. Ibid., p. 151.
10. Evelyn P. Foote, "Army Leadership: One Point of View," in James H. Buck and Lawrence J. Korb, eds., *Military Leadership*, (Beverly Hills, Calif.: Sage, 1981), p. 559.
11. Rogan, *Mixed Company*, p. 131.
12. Nicholas Senn, *Medico-Surgical Aspects of the Spanish-American War* (Chicago: Medical Book Company, 1901), pp. 318–319.
13. Philip Kalisch and Margaret Scobey, "Female Nurses in American Wars: Helplessness Suspended for the Duration," *Armed Forces and Society* 9, no. 2 (1983): 221.
14. Holm, *Women in the Military*, p. 109.
15. Carol Boyd Lean, "Working for Uncle Sam—A Look at Members of the Armed Forces," *Monthly Labor Review* 107, no. 7 (July 1984): 3.
16. Chongsoo Kim, et al., *The All-Volunteer Force: 1979 NLS Studies of Enlistment, Intention to Serve, and Intentions to Reenlist* (Columbus: Ohio State University, 1982).
17. Army recruiters emphasize that the army issues certificates of completion for certain kinds of training. These certificates are valuable because they are acceptable as proof of mastery by civilian organizations.

18. DOL, *Women at Work*, p. 17.
19. Jerald Bachman, "American High School Seniors View the Military: 1976–1982, *Armed Forces and Society* 10, no. 1 (Fall 1983): 86–104; Vonda L. Kiplinger, David P. Boesel, and Kyle T. Johnson, *Propensity of Young Women to Enlist in the Military: A Report to Congress* (March 1985); see, for example, J. Hicks, "Women in the U.S. Army," *Armed Forces and Society* 4, no. 4 (Summer 1978): 647–657; Kathleen Durning, "Attitudes of Enlisted Women and Men Toward the Navy," *Armed Forces and Society* 9, no. 1 (Fall 1982): 20–32; Patricia Thomas, "Enlisted Women in the Military," in *Defense Manpower Policy: Presentations from the 1976 Rand Conference on Defense Manpower* (1978).
20. Chongsoo Kim, *The All-Volunteer Force*, p. 25.
21. Ibid., p. 26.
22. Charles C. Moskos, "Female GIs in the Field: Report from Honduras," *Society* 22, no. 6 (1985): 29.
23. Patricia M. Shields, "Women as Military Leaders: Promises and Pitfalls." Presented at the National Conference of the American Society for Public Administration, Indianapolis, Ind., 1985, p. 13.
24. DOD statistics.
25. Herbert Parnes, *Peoplepower: Elements of Human Resource Policy* (Beverly Hills, Calif.: Sage, 1984), p. 111.
26. Linda J.Waite and Sue E. Berryman, *Women in Nontraditional Occupations: Choice and Turnover* (Santa Monica, Calif.: Rand, 1985), p. 49.
27. DOD, *Military Women*, p. 51.
28. These conclusions appear to contradict the findings of Waite and Berryman. After controlling for marriage, birth, and branch of service, they found no significant relationship between military job turnover and work in nontraditional jobs. This study, however, did not distinguish between dirty, outdoor nontraditional jobs that require heavy lifting and indoor nontraditional activities, such as electronics.
29. DOD, *Military Women*, p. 53.
30. Ibid., p. 59.
31. Kim. *All-Volunteer Force*, p. 103.
32. Ibid., p. 95.
33. Ibid., p. 103.
34. Chongsoo Kim, *Youth and Military Services: 1980 National Longitudinal Survey Studies of Enlistment, Intentions to Serve, Reenlistment, and Labor Market Experience of Veterans and Attriters* (Columbus: Center for Human Resource Research, The Ohio State University, 1982), p. 87. Since Kim's study, the armed forces have recognized and are dealing with the critical day-care issue in a more explicit manner.
35. Mary Geurts, *Career Intent of Female Officers at Fort Hood, Texas.* Applied Research Project, Southwest Texas State University, 1985, pp. (appendix) I:1–7. This study was performed by a woman army officer to fulfill a requirement for a master's degree. Because respondents were helping a fellow woman officer and knew the results would not be used for official army business, one would expect the answers to be honest. Unfortunately, the sample size was small. The random sample was drawn from Ft. Hood's list of officers. However, only 36 of the 75 who received the mailed questionnaires responded. This represents approximately 11 percent of Ft. Hood's women officers. The instrument was adapted from an Air Force Quality of Life survey.
36. Holm, *Women in the Military*, pp. 51–54.
37. See, for example, Rogan, *Mixed Company*, pp. 184–223; Janice D. Yoder, et al., "The Price of a Token," *Journal of Political and Military Sociology* 11 (Fall 1983): 325–337.
38. Michael Rustad, *Women in Khaki: The American Enlisted Women* (New York: Praeger, 1982), p. 135.
39. Ibid., p. 159.
40. Robert H. Baldwin and Thomas Daula, "The Cost of High-Quality Recruits," *Armed Forces and Society* 11, no. 1 (Fall 1984): 164.
41. Mady Wechler Segal, "Woman's Roles in the U.S. Armed Forces: An Evaluation of Evidence and Arguments for Policy Decisions," in Robert K. Fullinwider, ed., *Conscripts and Volunteers* (Totowa, N.J.: Roman & Allanheld, 1983), p. 211.
42. Frank Partlow, "Womanpower for a Superpower: The National Security Implications of Women in the United States Army," *World Affairs* 146, no. 4 (Spring 1984): 289.

43. M.C. Devilbiss, "Gender Integration and Unit Deployment: A Study of GI Jo," *Armed Forces and Society* 11, no. 4 (Summer 1985): 523–552; Moskos, "Female GIs."

44. Devilbiss, "Gender Integration." pp. 542–544.

45. Holm, *Women in the Military*, p. 383.

46. Kim, *All-Volunteer Force*, p. 87.

47. Patricia M. Shields, *Women in the Military*. Report submitted to the Southwest Texas State University Organized Research Committee, 1982, pp. 18–19. This study used the 1978 NLS data.

48. A.V. Bartlett, et al., "Public Health Considerations of Infectious Diseases in Child Day Care Centers," *The Journal of Pediatrics* 105, no. 5, (November 1984): 683–701.

49. Child care for active duty personnel was a key agenda item at a recent meeting of the Defense Advisory Committee on Women in the Services. Nonna Cheatham, "DACOWITS Holds Fall Meeting," *Minerva* 3, no. 1 (Spring 1985): 2.

VIII

Race Relations in the Military

JOHN SIBLEY BUTLER

The purpose of this chapter is to examine the institutional—occupational thesis from the standpoint of race relations. Since the publication of Charles Moskos's seminal article about 10 years ago, an entire research tradition has evolved. Although much of this research contributes to an understanding of the I/O thesis, my contention is that considerable confusion has developed about the thesis. Put differently, although Moskos's original conceptualization was grounded in the sociological tradition, which emphasizes structure as the unit of analysis, most of the empirical work utilizes the individual as the unit of analysis. In order to provide a full example of the Moskos thesis, race relations within the military will be utilized as data: an examination of substantive issues of race relations will inform our theoretical understanding of the I/O thesis.[1]

The I/O thesis represents a continued effort to capture the relationship between civil and military organizations. In a real sense, it is a natural outgrowth of the convergence—divergence theory that has influenced military sociology for the last four decades. This theoretical tradition examines the extent to which organizational similarities and differences occur in military and civilian structures. To give theoretical balance and structure to this thesis, Moskos suggested that the convergence—divergence thesis should concentrate on the extent to which similarities or dissimilarities exist between civil and military organizations as measured by organizational structure, skill levels, membership representation, and ideological components.[2] A host of related empirical work has been accomplished. General conclusions of this research stress the idea that some structures of the military will be convergent (i.e., technical, administrative, and clerical) and others will be divergent (i.e., combat in nature).[3]

The switch from a military based on conscription to one based on volunteers stands as the single, major event that spanned the I/O thesis. Instead of concentrating on measuring subtle changes of convergence or divergence, the creation of the all-volunteer force clearly called for a new theoretical conceptualization of the civilian-military interface. Moskos stepped into this theoretical vacuum to put forward his ideas.

He chose the concept of institution versus occupation in order to capture the change in the military, given the development of the all-volunteer force. Although all occupations are certainly institutions in the sociological sense (in that institutions are organized ways of doing things), these metaphors were designed to discriminate between *structural* features associated with different organizational formats. Such features are expected to exert a certain force on individuals that will result in their embracing certain values associated with the ability of people in groups to function as a unit. Recognizing that the concepts represent ideal types that are not necessarily mutually exclusive, Moskos developed their operational definitions.

An *institution* is legitimated by values and norms that transcend individual self-interest in favor of a presumed higher good. Features in the armed forces associated with this type of format include extended tours abroad, fixed terms of enlistment, liability for 24-hour service, inability to resign, strike, or negotiate working conditions, subsidized base facilities, and remuneration based on rank, seniority, and need.[4] Conversely, an *occupation* receives its legitimacy in terms of the marketplace, which emphasizes prevailing monetary rewards for equivalent competencies. Features associated with this format include a negotiation between individual and organizational needs because of the cash-work nexus, a priority of self-interest rather than that of the employing organization, trade unionism, and separation of work place and residence.[5] Using developmental analysis, which allows one to project future organizational forms based on the construction of a pure type, Moskos then shows that the logic that created the all-volunteer force will have the effect of switching the military from an institutional to an occupational format.

Moskos's analysis is structural, not based on the individual wishes of military personnel. We make this point because research attempting to measure Moskos's metaphor has utilized *individuals* as the unit of analysis.[6] Although this work adds to our knowledge of the military, it cannot be viewed as a direct measure of Moskos's ideas, which are *structural* in nature.

Central to Moskos's concept is that if you change certain structural features of an organization, individuals will respond in predictable ways. Thus, his "features"—outlined as relating to the institutional and occupational formats—are really structural variables. For the institutional format, noted above, features include inability to resign or negotiate working conditions, compensation in noncash forms, and subsidized consumer facilities. Structural features associated with the occupational format include the legitimation of the cash-nexus basis, separation of workplace and residence, fewer subsidized consumer facilities, and increased numbers of civilian employees.

Note that in Moskos's original ideas nothing appears about the attitudes of military personnel. He did say, however, that if a cash-nexus system is established, military personnel will inexorably begin to concentrate on increasing their pay, and military unionization becomes a possible vehicle for that goal. Likewise, if you separate workplace from residence, military personnel will react by becoming like their civilian counterparts—arriving early and leaving late in the afternoon to avoid commuter rush-hour traffic. Put simply, because of structural changes in the military, it begins to resemble an occupation rather than an institution. Its members in turn begin to act as if they were in an occupation rather than in an institution. These variables are over and above the individual and exert a pressure that impels them to respond in a certain way.[7]

An analysis of race relations in the military will illustrate the dynamics of the Moskos thesis. The changing *structure* of the military is what has affected race relations. Put another way, the I/O thesis places priority-both in importance and in chronology—on structural change. Alterations in attitudes and behavior follow structural change, not the other way around. This theme is especially clear in the abundant literature on race relations in the military.

THE I/O THESIS AND RACE RELATIONS

Structural changes accompanying the shift to the all-volunteer force are altering the pattern of race relations in the military. My basic hypothesis is that as the military becomes more occupational, race relations converge with those of civilian life. This proposition is tested through developmental analysis. This method, which entails developing a model that attempts to project the future state of an organization, is based on predicting outcomes given organizational trends and structural changes. It allows one to theorize about the future state of a specific factor within an organization.[8] Of first concern is the development of propositions derived from research on race relations and the military prior to inception of the all-volunteer force. Following that is a discussion on the development of theoretical propositions that should be expected vis-à-vis race relations under the present and future all-volunteer force. Propositions in both sections are grounded in historical events. My final comments address the measurement of the I/O thesis.

Historical Considerations

From this nation's very beginning, and even before its inception, blacks participated in military organizations. This long history of participation has provided a natural laboratory for the development of theoretical propositions about race and military organizations. Theoretical treatments begin by stressing the fact that Crispus Attucks, a runaway slave, had been the first to die during the Boston Massacre. When the Revolutionary War began, Gen.

George Washington's headquarters issued four orders forbidding black enlistment in the Continental Army. This action barred thousands of black slaves and freedmen who were willing to bear arms on behalf of the colonies. This exclusion was the decision of the Colonial Counsel of Generals, many of whom were from the plantation colonies. Early in the war, the British forces offered blacks their freedom if they would join the side of the Crown. When blacks began to enlist in the British forces, General Washington himself sent a letter to members of the Continental Congress, informing them that he would enlist blacks even if they did not agree. The Congress did not hinder the enlistments, and over 5,000 blacks served in the Revolutionary War.[9]

There were some instances of equal treatment in race-mixed units of the war, but the typical black soldier was a private, tending to lack official identity. Often his record bore no specific name. He was carried on the rolls as 'A negro by name," or "A negro name not known."[10] The seven brigades of Washington's army averaged 54 blacks each. When victory was assured, the ideology of war—with its promises of full, equal treatment within military organizations and civilian life—was replaced by the conservatism of the constitutional era. Blacks were excluded from participation within military institutions.

At the start of the Civil War, exclusion of blacks continued. When Northern blacks responded with enthusiasm to the first call for volunteers, the secretary of war said bluntly, "This department has no intention to call into service any colored soldiers."[11] Only after the Emancipation Proclamation were blacks finally allowed to enlist on the side of the Union. War Department order GO143 in 1863 allowed black enlistment, and eventually nearly 180,000 blacks fought on the Union side in 120 separate army units designated "United States Colored Troops." These segregated troops fought in pivotal battles, won 14 Congressional Medals of Honor, and played roles in the liberation of Petersburg and Richmond.[12] When the war ended, the War Department took the position that the enlistment of blacks was a "peculiarity of the volunteer service" that had not yet been authorized for the regular armed forces.[13]

After the Civil War, the Radical Republicans, a group that had played a part in the Reconstruction, argued that "blacks in blue" should be integrated into the regular military establishment of the United States. This recommendation was never followed; instead, the army created four black units—the Ninth and Tenth Cavalry and the Fourteenth and Fifteenth Infantry—that played a major role in the Indian wars from 1870 to 1890.

In June of 1898, blacks responded to the call to arms to fight on Cuban soil. One of the most publicized events of black military history was the bravery of those who joined Teddy Roosevelt as Rough Riders. After the charge up San Juan Hill, Roosevelt wrote, "I want no better men beside me in battle than these colored troops showed themselves to be."[14] In 1906, in Brownsville, Texas, black troops were accused of riding through the city, firing at whites.

Then Roosevelt, who once had nothing but praise for black soldiers, dismissed three companies of black troops with dishonorable discharges. Race relations declined in the military as well as in the civilian sector. In 1896, the *Plessy* vs. *Ferguson* decision created the term *separate but equal*, and there were antiblack riots and lynchings throughout the South.

Despite the decline in race relations, when the United States declared war on Germany in 1917, black leaders urged blacks to support President Wilson's call to "make the world safe for democracy." At the same time, Frank Park, a congressman from Georgia, introduced a bill to make it unlawful to appoint blacks to the rank of either noncommissioned or commissioned officer. During World War I, the army maintained traditional (segregated) black regiments. More than 200,000 served overseas, 150,000 in backbreaking stevedore battalions.[15]

During the period between World Wars I and II, the army remained segregated; it adopted a policy, the quota system, that kept the number of blacks in the army proportionate to their numbers in the total population. By 1940, there were only 5,000 men in all-black units and only five officers.

At the outbreak of World War II, America turned again to its black population for manpower. The segregated army again held sway, with old units reactivated and new ones formed to accommodate the more than 900,000 blacks who served. During this war, however, civil rights groups began to pressure Congress for changes in the structure of the military. A. Philip Randolph called for a march on Washington; Bayard Rustin and Elijah Muhammad dramatized their position by going to jail. In 1948, President Truman signed an Executive Order officially outlawing segregation in the armed forces. The Truman order had a significant effect on the Korean conflict (1950–53), with its need for manpower. By May 17, 1951, 20 percent of black soldiers in the Eighth Army were in integrated units, three-fifths of them in combat arms.[16]

Following the Korean War, the Cold War era saw continuing improvement in race relations. In Vietnam, race relations metamorphosed into a pattern of systematic conflict, although the official policy of the military was that of egalitarianism.

THE SOCIOLOGY OF MILITARY RACE RELATIONS

The massive historical and sociological data from the Revolutionary War through World War II have produced two major proven propositions: (1) although blacks have been involved in all conflicts, their experiences vis-à-vis race relations did not vary significantly from the systematic unequal treatment and segregation ongoing in the civilian sector. (2) After participation in military conflicts, blacks were rejected from the military almost totally when peace reemerged; this latter proposition is grounded in the recruitment, retention, and discharge experiences of blacks within the organization.[17]

Notwithstanding, this strong tradition of racial segregation was completely turned around when the movement toward racial egalitarianism began. Although issues of inequality would continue to be present within military institutions, the issue of desegregation rapidly became a nonissue. The decisive transformation in race relations can be accounted for by the unique features of the armed forces that were strongly in place at the beginning of desegregation—its rigid, hierarchical structure and its separation from the larger society.

Being a hierarchical structure means that the military is a highly crystallized organization: the distribution of authority comes from a central place and is implemented through a chain of command. Each office, or rank, has specific power and privileges. Especially unique to military organizations is the presidential nexus. The president is not only the chief executive of the country, he is also commander in chief of the military. Any executive presidential order flows directly to the military without question.

Tied to this hierarchy are the numerous systems of protocol and justice that legitimate the structure. Individual members are aware that each step up the rank structure represents prestige, power, and authority. Because of this highly crystallized arrangement, once the president gave the order for desegregation, things moved rapidly. Put another way, because the hierarchical structure is patterned on stable and predictable relationships, decisions regarding race did not have to take into account the personal desires of military personnel. If the military had been a democratic civilian institution, then desegregation would have been problematic as it has been in civilian institutions.

Also contributing to the transformation of race relations was the time in which it took place; integration occurred during a period of conscription. This meant that an element of civic duty was ingrained in military institutions at the same time that integration was taking place. During the Korean conflict, conscription brought a cross-section of American youth into the military. Thus, the "rapid" integration should also be viewed in light of the "forcing" together of people with divergent backgrounds due to conscription.[18] During this period of rapid transformation, there was almost a complete separation of workplace and residence; the military in essence represented a separate entity. All primary needs of personnel were met. This separation, especially during training periods, was done in order to instruct individuals about military organizations and to help them develop military skills. Once the pattern of racial desegregation started, the discontinuous nature of military life enhanced the rapid transformation of race relations. The combination of workplace and residence meant that blacks and whites were forced to live with one another. This forced racial contact resulted in more positive racial attitudes between the two groups.

The classic World War II studies of Samuel Stouffer and associates produced one of the most significant sociological propositions on race relations—the *contact hypothesis*. The proposition held that under certain conditions, the more contact individuals from different races have with each other, the more positive their attitudes toward each other. The conditions under which this proposition operates include (a) when an authority strongly sanctions interaction, (b) when there are commonly shared goals, (c) when the contact is by equal-status individuals, and (d) when interaction between individuals is cooperative and prolonged, covering a wide range of activities.

The military organization provided the unique setting that encompassed all the conditions. *The American Soldier* data asked the following: "Some Army divisions have companies which include Negro platoons and white platoons. How would you feel about it if your outfit was set up something like that?" The findings were that men actually in a company containing a black platoon were most favorable toward it; those in larger units with no mixed companies in a regiment, or in a division containing mixed companies held "intermediate" opinions. Put simply, the closer men approached the mixed-company status, the more they favored serving in a racially mixed organization.[19] Since the publication of *The American Soldier*, the contact hypothesis has stayed intact within military institutions: the more racial contact, the less negative attitudes.[20]

In addition to attitudinal change, research began to show that black soldiers perceived the military as being more equalitarian than the civilian sector. Although the organization was not without racial problems, they saw better opportunities for advancement and economic stability than in the civilian sector.[21]

The variables enumerated as responsible for the rapid transformation in race relations within the military are structural. Combination of workplace and residence, separation from the larger society, and military structure are variables that do not take into consideration individuals' attitudes. Indeed, they produce the attitudes and build race relations. Such an understanding brings us closer to the Moskos thesis. Just as structural variables had an impact on race relations prior to the all-volunteer force, Moskos is arguing that they will also have an impact on the current structure of the all-volunteer force. Let us now apply the Moskos model to race relations within this force to show how changing important structural variables should alter race relations.

To use race relations as data does not mean that the changes associated with the switch from an institutional to an occupational format will impact on only race relations. Because of this, as we delineate structural changes in the organization, we will discuss nonracial situations when necessary.

In a real sense, Moskos's conceptualization revolves around the military as both a garrison and a nongarrison entity. One of the unusual characteristics of a garrison military is the combination of workplace and residence. Thus, it is not surprising that Moskos notes:

> A hallmark of the conventional garrison Army has been the adjacency of work place and living quarters. As late as the mid-1960s, it was practically unheard of for a bachelor enlisted man

to live off-post. Not only was it against regulations, but no one could afford a private rental on a draftee's pay. Today, although precise data are not available, a reasonable estimate would be that as many as 30% of single enlisted people in statewide posts have apartments away from the military installation.[22]

One would expect that a structural change to the occupational model would have a significant impact on relations among the troops in general. The closeness, the sharing of different cultures, and the development of close, lasting friendships that developed in the garrison military should, by definition, decline. The hallmark of barracks life was that people from different socioeconomic backgrounds and cultures were forced to develop a method of adjusting to each other. Although not as yet verified, impressionistic evidence is very strong that whites are much more likely to live off post than blacks. Thus, under the all-volunteer force the barracks became more "black" at night than is the unit during the day. Such a pattern converges with civilian patterns of work and residence. Put differently, the impressionistic pattern of military life is somewhat analogous to contemporary urban life: the downtown area becomes black at night as whites return to the suburbs.

As noted in the previous section, the pre-all-volunteer military produced a forced type of racial contact. The separation of workplace and residence should decrease and alter this contact. As a result, the friendships and camaraderie that developed as a result of different races living together should decline. Racial contact thus will be more like the superficial exposure experienced by the majority of people in the general society. Spontaneous recreational events (going bowling or out to see a movie, etc.) should also decline. In the framework of the I/O thesis, race relations within the military will begin to resemble race relations within the civilian sector; such changes will accelerate if different races return to segregated neighborhoods after duty hours. Even if they do not, the close contact associated with the institutional format can never be accomplished under the occupational format. Initially, this projected change in race relations has nothing to do with individual attitudes. The structural change that takes place in the relationship between the workplace and residence is the explanatory variable.

Another structural variable that Moskos associates with the occupational format is the increase in the number of contract civilians. From 1964 to 1978, contract civilians rose from 5.4 percent to 14.8 percent as a proportion of total defense manpower. These civilians perform tasks that range from routine housekeeping and kitchen duties to quasi-combat jobs, such as manning missile warning systems and doing basic repair and assembly of hardware.[23] One could project that as the number of contract civilians increases, the less "military" (institutional) a job becomes. This presupposes that military and civilian personnel will work together on certain tasks. Without pay parity between them, one could expect that both the morale and job satisfaction of military workers will be affected.

This structural change could affect race relations within the military, but only if the following conditions exist: the civilian work force would have to be heavily composed of majority group members; these individuals would have to possess a history of working in a one-race environment; and (because of a former segregated workplace) they would have to hold negative racial attitudes toward the participation of blacks in the workplace. If these conditions are met, the positive race relations generated by military personnel in the workplace, instituted by command, could be adversely affected: the work environment in some military settings will mirror certain civilian workplaces.

Consider the following, highly plausible situation. Highly technical civilian contractors (e.g., "technical representatives" aboard ships), those who perform high-level maintenance in such areas as ordnance are almost exclusively white, but in the contract work for kitchen and janitorial work, civilians are disproportionately black. The point is that the race stratification underlying civilian life is replicated in the contract civilian force of the defense establishment. This is in contrast to the military side of the establishment, where blacks are heavily in noncommissioned grades and are more scattered throughout all occupations; this also holds somewhat for black officers. Put another way, civilian contractors within the military create the analog of the racial stratification system of civilian life.

The juxtaposition of white civilians, who may have little or no history of working with blacks, also could create racial tensions in the workplace similar to those found in the civilian sector. This would be new to the military, because the racial conflict during the 1960s had more to do with cultural variables and ideology (e.g., political ideas, Afro haircuts) than the workplace per se.

The percentage of blacks within the service will also have an effect on race relations. Although scholars have debated extensively the effect of the number of blacks in the military, it is plausible to theorize about how the black percentage issue in the all-volunteer force affects general race relations.[24] Moskos did not include this variable in his outcomes associated with the occupational outcomes of an organizational format, but it must be considered.

The first thing to note is that racial distribution, too, is a structural variable. Thus, we expect it to affect general attitudes and actions of individuals and groups. Sociological literature has a rich tradition of looking at the black percentage effect on general outcomes in society. We know, for example, that as the black percentage of a neighborhood increases, white commitment to that community decreases. The same is true for schools and—to an extent—occupations.[25]

Butler and Holmes theorize on how the percentage of blacks within the military affect it.[26] They argue that if we assume a convergence between the civilian and the military, the following is expected within the ranks. Whites may gravitate toward military occupations that do not have a high percentage of blacks. Thus, while the organization retains the latter, they will tend to be located only in occupations in which they are the majority. This is true to an extent today; the percentage of blacks is high in some assignments, such as in supply, and low in others. The projected effect of the black percentage in the military should be examined as the all-volunteer force continues.

Closely related to black percentage is the effect of the so-called *class* factor, given the present all-volunteer force. For example, the absence of the typical white middle-class individual from the enlisted ranks should have an impact on the quality of race relations within the military. Although the strata of whites who join the military vary according to the strength of the civilian economy, one can still say that middle-class whites are less likely to be in the ranks. Moskos, in an earlier work, commented that the all-volunteer force has attracted the "worst of the whites."[27] Because of a number of circumstances in American society, including competition in the labor force, conflicts over desegregation of public schools, and competition for inner-city housing, there has been systemic conflict between lower-class whites and minority groups (the black middle class is more likely to live in suburbia).[28] Interacting with this is the fact that the military is drawing a high proportion of whites from rural areas. Because of the class levels of whites, given the race-class dynamics of the civilian sector, one should expect class-race conflict to develop in the military of the future. Under the occupational format, the military structure can mitigate, but is less likely to prevent, this conflict.

In a real sense, it seems that the more occupational the military becomes, the more race relations worsen. In a major research effort, Butler and Holmes[29] found that blacks' perceptions of racial discrimination varied by service branch. What is interesting, given the Moskos I/O ideas, is that blacks in the marines perceive considerably less discrimination, on average, than those in the other three services. The authors attributed this finding to the influence of the marines' traditional structure of group solidarity and commitment to the corps. Put differently, the Marine Corps, which has largely retained the traditional institutional format, is seen in the best light by blacks vis-à-vis discrimination. Thus, the variation in service structure explains perceptions of discrimination.

Another interesting point about military institutions is that because pay is based on rank and not skill, there is racial parity. Overall, however, to the extent that an occupational format stresses the idea of remuneration based on skill, then by definition, pay inequalities will develop. This development would have to take place within the rank structure; thus, two members of the same rank could receive different incomes. In terms of race, to the extent that skill differences exist between blacks and whites, then the pay differentials would be by race. Such an arrangement would resemble the pay structure of the civilian sector (if—and only if—whites bring more skills to the military).

This means that the military would lose the distinction of being the only institution in America in which both races share the exact same pay structure.

No theorizing on the all-volunteer force would be complete without a consideration of the relationship between the military and the political institutions of the civilian sector. The general field of civil-military relations examines the interplay between soldiers and politicians, showing how military rule develops or is inhibited.

The development of the all-volunteer military moved away from the concept of the citizen-soldier. Although some have argued that such a concept is romantic, for much of our history, citizens rotated from the military to civilian society and vice versa. One of the classic propositions of civil-military relations is that the more citizens (rather than professional soldiers) are involved in the military, the higher the probability of civilian control. Conversely, when the military develops a "professional" motif, the possibility of military rule is enhanced. The rotation of citizenry (back and forth between the military and civilian society) leads the citizen to develop self-interest in an established civilian government.

The all-volunteer force should add an interesting twist to civil-military relations in America. First, we will rule out the actual probability of a military form of government in this country. However, because Congress makes decisions about military pay, benefits, and the like, we may see military personnel becoming much more responsive to the wishes of Congress. The somewhat adversarial relationship between the military and Congress over the military budget, for example, may be softened. This could happen because military personnel, given the occupational format, would have to bargain to enhance or retain the perks of the military. Put differently, given the occupational format, Congress becomes more of a management structure. Such an analogy would have to become stronger if the all-volunteer force is to continue within the occupational format. In an interesting way, those in the military would have to pay much more attention to this management structure (Congress). Before the inception of the all-volunteer force, Congress functioned simply as an entity responsible for "maintaining and raising an army." Now Congress must also keep that army satisfied—because of its volunteer status.

In terms of race relations, this means that black soldiers, like their white counterparts, will be more likely to pay attention to civilian leaders. Under an institutional format, without concerns about salary increases, job satisfaction, working conditions, and other such matters, soldiers were not concerned with Congress. During the recent U.S. interventions in Grenada and Lebanon, Jesse Jackson noted that blacks were being used as "cannon-fodder." The Congressional Black Caucus also came out against the intervention in Grenada. To the extent that the military is institutional, soldiers would be less likely to give much consideration to such statements by political leaders. (This would be true also for all ethnic groups in the country.)

CONCLUSION

Using race relations as data, this chapter has tried to show that Moskos's I/O theoretical reasoning is grounded in a structural approach. This was needed because more empirical approaches to the Moskos thesis utilize individual attitudes as data. Although such studies are important, they ultimately do not measure his I/O model. Some final comments on how this can be done are appropriate.

In order to build a research tradition that measures the I/O thesis, individual researchers and those associated with the Department of Defense need to collect data that allow one to create structural variables. For example, we need to know the actual numbers of individuals who live on and off military bases. Because of variation by base, researchers may create propositions, such as the more separation of workplace and residence, the lower group solidarity of a base, company, platoon, and so on; the higher the incidence of workplace and residence separation, the less commitment of military personnel. Notice here that the variation of how individuals relate to the military revolves around a structural variable, the separation of workplace and residence. The same approach could be utilized by creating an index of military bases *vis-à-vis* the number of civilian employees, the proximity of civilian unions to the base, the relationship of the service spouse to the military community, vertical or horizontal reference groups, and other structural variables developed in Moskos's work.

The development of such measures should allow us to continue to assess the relationship between the changing structure of the military and race relations. Our major thesis has been that the occupational format of the all-volunteer force serves to alter the pattern of race relations that was established before inception of the all-volunteer force. More to the point, the pattern of race relations within the military gravitates toward that found in the civilian sector, making the military a reflection of the latter. Just as the organizational structure helped to account for the rapid change in race relations in the conscript military toward equalitarianism, it will likewise alter race relations in the all-volunteer force.

NOTES

1. The seminal paper that addresses this thesis is John Sibley Butler and Malcolm D. Holmes's "Changing Organizational Structure and the Future of Race Relations," in Robert K. Fullinwider, ed., *Conscripts and Volunteers* (Totowa, N.J.: Rowman & Allanheld, 1983), pp. 167–177.
2. Charles C. Moskos, "The Emergent Military: Civil, Traditional, or Plural?" *Pacific Sociological Review* 16, no. 2, (April 1973): 255–279.
3. See Morris Janowitz, *The Professional Soldier* (New York: Free Press, 1960); Charles C. Moskos, *The American Enlisted Man* (New York:Russell Sage Foundation, 1970); David Segal, John Blair, and S. Stephens, "Convergence, Isomorphism and Interdependence at the Civil-Military Interface," *Journal of Political and Military Sociology* 2: 157–172.

4. Charles C. Moskos, "The Enlisted Ranks in the All-Volunteer Force, in John B. Keeley, ed., *The All-Volunteer Force* (Charlottesville, Va.: University of Virginia Press, 1976), pp. 53–56.

5. Ibid.

6. See, for example, Charles A. Cotton, "Institutional and Occupational Values in Canada's Army," *Armed Forces & Society* 3 (1981): 99–110; John Sibley Butler and Juanita Firestone, "Institutional and Occupational Orientation: Implications for Organizational Commitment." Unpublished manuscript, Department of Sociology, the University of Texas, Austin; David R. Segal and Young Hee Yoon, "Institutional and Occupational Models of the Army in the Career Force," *Journal of Political and Military Sociology* 12 (1984): 243–256.

7. Structuralism in sociology can be traced to Emile Durkheim. In order to separate sociology from psychology, he proposed that the correct unit of analysis for sociology was the social fact. Social facts are variables that cannot be associated with individual consciousness, but stand over and above the individual. They include norms, institutions, values, and sanctions. Human behavior is explained by examining how these factors exert different kinds of pressures on individuals. See Emile Durkheim, *The Rules of Sociological Method* (New York: Free Press, 1938).

8. This method has been utilized extensively in the literature in organizations. As far as military organizations are concerned, see Harold Lasswell, "The Garrison State" *The American Journal of Sociology* 46, no. 3 (1941): 455–468. Charles C. Moskos also utilized the method in his original piece "From Institution to Occupation," *Armed Forces & Society* 4 (Fall 1971): 41–50.

9. John Sibley Butler, *Inequality in the Military: The Black Experience* (Saratoga, Calif.: Century Twenty One Publishing, 1980), pp. 19–21.

10. "Lists and Returns of Connecticut Men in the American Revolutionary War, 1715–1782," Connecticut Historical Society, Coll., 12 (1909), 59, 80, 81, 183.

11. Richard J. Stillman, *Integration of the Negro in the U.S. Armed Forces* (New York: Frederick Praeger, 1968), p. 9.

12. Dwight W. Hoover, *Understanding Negro History* (Chicago: Quadrangle Books, 1968), p. 270.

13. John D. Foner, *The United States Soldier: Between Two Wars* (New York: Humanities Press, 1970), p. 127.

14. Herschel Cashin, *Under Fire* (New York: Tennyson Neeley, 1899), p. 147.

15. John P. Davis, *The Negro Reference Book* (Englewood Cliffs, N.J.: Prentice Hall, 1976).

16. Butler, *Inequality in the Military*, pp. 26–28.

17. For further discussion, see Butler and Holmes, "Changing Organizational Structure."

18. Charles C. Moskos, "Blacks in the Army," *Atlantic Magazine*, May 1986, pp. 69–72.

19. Samuel Stouffer, et. al., *The American Soldier* (Princeton, N.J.: Princeton University Press, 1949), vol. 1, p. 594.

20. Butler and Holmes, "Changing Organizational Structure." For research on the contact hypothesis, see Stouffer, et. al., *The American Soldier*; John Sibley Butler and Kenneth L. Wilson, "The American Soldier Revisited: Race and the Military," *Social Science Quarterly* 59 (December 1978): 451–467.

21. See, for example, William J. Brink and Louis Harris, *Black and White: A Study of U.S. Racial Attitudes Today* (New York: Simon & Schuster, 1967).

22. Moskos, "The Enlisted Ranks," p. 58.

23. Ibid., p. 60.

24. See, for example, Morris Janowitz and Charles C. Moskos, "Racial Composition in the All-Volunteer Force," *Armed Forces & Society* 1 (Fall 1974): 109–123; and Alvin J. Schexnider and John Sibley Butler, "Race and the All-Volunteer System," *Armed Forces and Society* 2 (May 1976): 431–432.

25. Butler and Holmes, "Changing Organizational Structure"; see also Michael W. Giles, Everett Cataldo, and Douglas S. Gatlin, "White Flight and Percent Black: The Tipping Point Reexamined," *Social Science Quarterly* 56 (1975): 85–92.

26. Butler and Holmes, "Changing Organizational Structure."

27. Moskos, "The Enlisted Ranks."

28. For a discussion, see William J. Wilson, *The Declining Significance of Race* (Chicago: University of Chicago Press, 1981).

29. John Sibley Butler and Malcolm D. Holmes, "Perceived Discrimination and the Military Experience," *Journal of Political and Military Sociology* 9 (Spring 1981): 17–30.

IX

Value Formation at the Air Force Academy

THOMAS M. McCLOY AND WILLIAM H. CLOVER

Moskos conceptualized the military as moving from an institutional perspective (normative values, duty, honor and country, service before self) to an occupational orientation (marketplace economy, contractual relationships, monetary rewards for equivalent competencies).[1] This chapter attempts to show the implications of this hypothesis at the United States Air Force Academy. All our data deal directly with Air Force Academy cadets, but we believe that this information can be generalized somewhat to other military academies.

The Air Force Academy is in the business of teaching and attempting to strengthen institutional values. The academy is not unique, however, in its attempt to teach the overarching values of the institution. Gary Dessler reports on research by William M. Mercer, Inc., that indicates that over 48 percent of surveyed companies pay a great deal of attention to the issue of corporate values.[2] These corporations are concerned with values such as performance, fairness, competitiveness, and team spirit. The Air Force Academy and other military academies differ from corporations in that they attempt to teach a single set of values to which they want their students to aspire. Corporations do not try to articulate service before self, patriotism, duty, honor, and country; however, such values provide the teaching focus of a military academy. In this regard, military academies are like religious schools.

Discussion of these issues at the Air Force Academy is appropriate because many people assume that a military academy is a place of purely institutional values. Hence, it is the most idealized institution within the larger military organization.

We will make several assumptions in this chapter. First, we believe that the internalization of institutional values is a developmental process that requires time to reach fruition. Second, we believe that development of institutional and occupational values can proceed simultaneously. A corollary to this assumption is that occupational values are desirable. From the standpoint of the organization, individuals possessing strong orientations toward achievement, competition, productivity, and success are absolutely essential for accomplishing missions in today's complex technical environment. The task of the organization is not to displace these occupational values, but rather to subjugate them to overarching institutional values when conflict arises between the two. Our third assumption is that institutional values are much less concrete than those we call *occupational*. Finally, we believe that a great deal of overlap exists between the construct of occupational values and the idea of *need for achievement* articulated by David McClelland.[3]

The idea that need for achievement confounds or overlaps with the concept of occupationalism is important to our understanding of cadets and the process that takes place at the Air Force Academy. Individuals with high achievement motivation have been characterized as having at least three consistent characteristics.[4] First, high-need achievers are individual goal setters, They like the sense of individual responsibility. Second, they like tasks that they assess as moderately difficult. They don't like tasks that are too easy because such tasks carry no sense of accomplishment; they don't like tasks that are too hard because success might be attributable to luck and not to their efforts. Third (and perhaps most important to our purpose in understanding the role of the occupational characteristics of the Air Force Academy), high-need achievers like tasks that provide immediate feedback concerning success or failure. We believe that aspects of the academy that many people see as occupational are actually the tangible rewards that high-need achievers must attain on the road to satisfying their long-range values. We agree that "man does not live by bread alone," but man does need bread to live as he or she strives to live according to, or up to, the institutional values for which the organization exists. Our high-need achievers at the Air Force Academy need the immediate feedback and success that many of the occupational and tangible aspects of the academy can provide.

THE ACADEMY SETTING

With these considerations in mind, we will attempt to clarify some aspects of the institutional-occupational hypothesis as it relates to the Air Force Academy. In the remaining portions of the chapter, we will discuss the Air Force Academy from several points of view. First, we will look at the recruitment process and the image that the Air Force Academy projects to high school students as they consider applying to the academy. Second, we will consider the selection process; how the academy winnows the large group of students who seek admittance. Third, we will discuss some of the reasons why cadets tell us they come to the Air Force Academy. These reasons will be based on data gathered from arriving cadets and again from cadets who have spent one to four years at the academy. Fourth, we will review how the academy affects cadets as they encounter the three major mission elements of the academy: military training, academics, and athletics. These elements, acting individually and in concert, shape the cadets' values. Fifth and last, we will discuss some of the data we gathered as we attempted to measure the institutional and occupational orientations of cadets. We will examine these orientations and try to show their relationship to important outcomes of the Air Force Academy experience, training, socialization, and education.

Recruitment

The marketing literature produced by the academy admissions office targets both the institutional and the occupational values of prospective cadets. From an institutional standpoint, the recruitment literature and process emphasize the role of the academy in developing career professional officers to lead others and, if necessary, to defend their country. Recruitment emphasizes the importance of commitment, selflessness, honor, and ethics, as well as the many military honors (including the Medal of Honor) that graduates have earned.

The occupational approach generally focuses on four areas: three aspects of opportunities at the academy as well as the availability of jobs upon graduation. In the first area, academics, the academy ranks among the best schools in the nation; it offers 23 academic majors, and several graduates have won prestigious scholarships, such as the Rhodes. A second area of occupational opportunities is athletics, an integral part of cadet life. "Not all athletes become Air Force Academy cadets, but all Air Force Academy cadets become athletes."[5] Every cadet participates either in an intramural program that involves 18 different sports or as a varsity athlete on one of 41 intercollegiate teams. The third area focuses on the unique opportunities offered to all cadets, such as parachuting or flying sailplanes and single-engine aircraft for those who are going on to pilot training. The final area concerns the future opportunities for graduates. The future is guaranteed: 60 to 70 percent of the graduates will enter pilot training, and others will enter fields such as engineering or space operations.

Selection

Every training program begins with selection. Once the organization has articulated criteria for selection it attempts to find individuals who meet the requirements. For those organizations that desire both institutional and occupational orientations, the problem of measurement arises. Inferring

TABLE 9-1. Reasons and Objectives Considered by Entering Students to be Essential or Important*

		AFA	Public Colleges
1.	Be well off financially	69.6%	71.7%
2.	Raise a family	77.4%	88.5%
3.	Keep up with political affairs	70.4%	38.0%
4.	Has a good academic reputation	90.1%	47.2%
5.	Offers financial assistance	53.8%	21.0%
6.	Graduates get good jobs	86.5%	43.6%
7.	Graduates go to top graduate schools	48.1%	20.7%
8.	Get a better job	64.6%	76.0%
9.	Prepare for graduate school	77.4%	88.5%
10.	Make more money	58.3%	67.8%

*Data provided by Institutional Research Directorate, United States Air Force Academy, 1985.

occupational propensities from statements citing the primary reasons for going to college, such as earning more money or getting a better job, is relatively easy (see Tables 9-2 and 9-3), but pinpointing institutional values is far more difficult. The academy attempts to assess this dimension through the candidate's past participation in leadership, group, or team activities. Being a Boy Scout, a member of an athletic team, or a high school class officer are all thought to indicate an orientation that values service and emphasizes teamwork over self-interest. Although selecting according to a quantifiable occupational measure, such as high school grade point average, is obviously easier, the academy attempts to include some institutional assessment as well. Although the above-mentioned activities may not measure institutional values in the purest sense, they are believed to reflect a predisposition toward institutionalism that can be nurtured and developed through later training and experience at the academy.

Reasons for Entering

Tables 9-1 through 9-3 give some of the reasons why cadets choose to come to the Air Force Academy. These reponses reflect answers that were given immediately upon entry and retrospectively after the cadets had been at the academy from one to four years. They suffer some limitations, but we have included them to show the consistencies and inconsistencies that exist in currently available data. Tables 9-1, 9-2, and 9-3s how the difficulties inherent in any attempt at purely institutional or occupational classification. When we compare entering Air Force Academy cadets with entering students at four-year public colleges, we note some striking similarities and differences. Lines 1, 2, and 3 of Table 9-1 suggest that all students have a desire to be well off financially, but Air Force Academy cadets state less of a desire to "have a family" and a much greater desire to "keep up with political affairs."

TABLE 9-2. Reasons for Coming to USAFA: Percentage of Cadets Selecting Various "Reasons for Coming to USAFA" by Sex and by Class*

		Freshman		Sophomore		Junior		Senior	
		M	F	M	F	M	F	M	F
1.	Be an officer	26	31	24	29	28	47	28	25
2.	Fly	32	19	33	11	37	12	38	11
3.	Good education	13	14	16	20	10	11	11	16
4.	Athletics	2	0	1	0	1	0	1	1
5.	Free education	4	5	5	4	4	3	2	4
6.	Find spouse	0	0	0	0	0	0	0	0
7.	Outside pressure	1	1	1	2	1	2	1	3
8.	Total challenge	15	24	12	24	9	18	10	24
9.	Other	8	6	9	10	8	8	7	14
	SAMPLE	827	120	641	91	624	66	699	91

*Data collected as part of survey of the cadet wing, March 1984.

TABLE 9-3. Principal Reasons for Coming to USAFA*

		1975	1983
1.	Free education	13.1%	10.7%
2.	High level of education	26.8%	28.5%
3.	AF career	19.2%	15.2%
4.	Fly	31.5%	33.3%
5.	Parents/friends	N/A	7.9%
	Sample Size	574	597

*Data collected systematically in political science classes at the academy in the "Kaydet Survey."

Table 9-1 can be interpreted as concern with occupational issues. On these dimensions, cadets score much higher than entering men and women at four-year public colleges. Issues such as good academic reputation, financial assistance, getting good jobs, and going to top graduate schools may seem to be occupationally oriented, but they may also represent a high need for achievement on the part of air force cadets. These concerns easily fit into the life patterns that cadets established in high school. Typically, cadets score in the top 10 to 15 per cent on the Scholastic Aptitude Test (SAT), and they typically have participated in more athletic and leadership activities than most high school students.

Lines 8 to 10 of Table 9-1 lead us to believe that the data in lines 4 to 7 have institutional meaning: entering Air Force Academy cadets cite such reasons as "getting a better job," "preparing for graduate school," and "making more money" at a lower rate than entering college students. The two most occupational items in Table 9-1 ("better job" and "making more money") are selected less often by cadets as reasons for going to college.

Although we argue from these data that cadets seem more institutional than other students, we recognize that they have their share of occupational concerns. We also believe that cadets' need for achievement may be indicated in the responses that appear in all three of these tables. The overlap between occupational orientation and the need for achievement may confuse our picture of the cadets' values, but we think that the data presented later will clarify the picture somewhat.

> Ever since I've been in the seventh grade I have wanted to be a cadet at the Air Force Academy ... now that I'm here all you people talk about is being an officer. I've never thought about that! (4th Class Cadet, Class of 1986).

As shown by this quote and in Table 9-3, the reasons for coming to the academy are many and varied. We would expect that cadets who are institutionally motivated would come to the academy primarily to become officers, and those who are occupationally motivated would come for other reasons, such as education. As shown in Table 9-2, only a fourth to a third of cadets, by their own admission, entered the academy to become officers. More than a third of the male cadets said they came to the academy to fly. This may seem to be an occupational orientation, but the following quote from James Webb's novel should give us pause in that automatic assessment.

> This whole business of flying jets was more than his job; it was the way he defined himself, and to interfere with that would be to destroy him. (*A Country Such As This*, p. 182.)

Knowing what a given activity means to any individual is not always easy; in many cases perhaps only the individual can decide whether an activity is occupational or institutional. Flying, for example, may be a manifestation of pure occupationalism, or it may represent something much more important and "institutional" ("It's the way I serve my country".)

Several points in Table 9-2 deserve considerations. First is the difference between men and women; women typically select "be an officer" in three out of four classes at a higher rate than men. Women also select "fly" at a much lower rate, but this finding is biased because fewer women than men are pilot-qualified (meeting necessary visual requirements, for example). Also, nearly double the percentage of women select "total challenge." It seems to us that women who attend this previously all-male academy have a high need for achievement. Finally, we note relative stability in the reasons cited and in their proportions throughout these tables. The pattern of responses that measure cadets' responses at different times show a general consistency.

Another source of data comes from systematic surveys collected by the Department of Political Science in upper-divisional political science classes. Here we see first that the reasons chosen for coming to the Air Force Academy remained fairly constant over this eight-year period. In addition, this survey, which distinguishes "free education" from "high level of education," gives a different picture of the relative percentages of cadets who appear to have an occupational reason for attending the Academy ("free education," 13.1 percent and 10.7 percent, respectively) and those who cite a more institutional, value-oriented reason "high level of education." The percentage choosing "to fly" in this sample is relatively consistent across the two samples and time periods.

WHAT THE ACADEMY DOES

Cadets entering the academy are typically young, intelligent, idealistic people with high needs for achievement. They also are disrupted and somewhat displaced from their roots; therefore, they seek new support structures. They were successful in the past and they stood out above their peers, but when they enter the academy, they are thrown into a group of individuals in which everyone is a top performer. Where do they turn to find that identity (as top student, athlete, class president) that previously provided stability and direction? The answer is this new organization, which provides new sources of identity, many of which are occupational, while institutional development occurs.

Throughout the four years of academy training, a dualism of institutional and occupational values develops in parallel fashion. The basic premise is that binding to the institution takes time, It probably doesn't occur until the graduate is in the air force, experiencing the real social exchange between the individual and the institution (e.g., personal commitment to the mission and institutional recognition in the reward). Perhaps a totally institutionalized state of mind is similar to "getting religion." In this situation, self-sacrifice for the good of the institution is legitimized by buying into the beliefs and the value system of the institution. This step cannot occur, however, until the individual has had sufficient exposure to the articulated belief system to develop an understanding and an appreciation for it. Such development occurs at the academy throughout the cadets' entire education and training process.

The education and training of cadets are accomplished primarily through the three mission elements of the academy: military, academic, and athletic. Although their functional areas are largely independent, the combined purpose of these elements is to "produce officers who exemplify leadership ability, strong character and a sense of duty, excellent physical fitness, sound intellectual ability, and above all motivation toward service to country."[6] This developmental process takes place over a four-year period, during which all the mission elements simultaneously make demands on the cadet.

Before discussing the impact of the three mission elements individually, we should address several general characteristics of the training process. First, the total immersion in institutional values is a central feature of the overall socialization process. The typical cadet day starts before 0600, finishes after 2300, and is filled with exposure to all three elements. Second, the faculty and the staff are almost entirely active-duty military personnel, primarily air force. Therefore, they occupy the position of linking pins between the cadets and the profession of arms. Insofar as faculty and staff have the "right stuff," it will be caught by the cadets as well as taught to them. These active-duty officers help provide the big picture and clarify the relationship between occupational and institutional value formation. For example, the military faculty will make it clear to a student studying engineering that his or her role upon graduation will be that of an officer with engineering expertise rather than that of an engineer who happens to be in the military. Finally, the process of institutionalism may be likened to immunization, in which occupationalism is the foreign body. To become fully immune to the ill effects of the toxin, the body must receive a series of inoculations. During the four-year training, cadets are exposed to repeated small doses of occupational experience in a controlled environment. The controlling influence might be the presence of the active-duty officer (an antibody?), who translates the current occupational experience into the long-range institutional perspective. In the long run, this process would allow institutional values (the individual's own immune system) to prevail in the presence of occupational inclinations, just as the body would eventually be able to function effectively in the presence of previously toxic foreign bodies.

Military Training

The summer before their first year at the academy, cadets experience Basic Cadet Training, a rigorous six-week program. During this period, they undergo the most intensive institutional indoctrination in their four years of academy training: they are deindividualized and then reshaped or molded within the institutional framework. The daily regimen makes considerable physical, mental, and emotional demands; the previous support structure of family, friends, and achievements vanish or at least are put at a distance, and new support structures that can assuage the stressful conditions must be found. The academy provides this support in the form of classmates; cadets learn quickly that cooperation enhances performance and mitigates stress. The lesson is clear: trying to scale a mountain alone is incredibly difficult, but working together as a team makes all obstacles surmountable and ensures survival.

During Basic Cadet Training, cadets learn the rules, regulations, and procedures necessary for everyday life at the academy. They receive honor and ethics training and are immersed in military history and culture as well as cadet rituals. This education lays the foundation for bonding to the institution, but the most significant bonding that occurs at this time is between classmates. A major goal of Basic Cadet Training is the establishment of class unity, the first step in the inoculation process. Cadets are encouraged to do their best individually, but to sacrifice individual performance if necessary by helping others so that they complete the training program successfully as a unit.

Successful completion of Basic Cadet Training results in a rite of passage called *acceptance,* the next significant step toward institutional bonding. First, basic cadets bonded with their classmates; now they officially become members of the Cadet Wing as fourth classmen. Although their primary loyalty and identification probably remain with their class, they now take on an added identity as members of the Air Force Academy Cadet Wing. The new identity is particularly evident in interactions outside the academy, which tend to strengthen it.

The next phase of military training is the fourth-class system, which begins after Basic Cadet Training and ends at the close of the spring semester of the first year at the academy. Training during this time focuses on the acquisition of professional knowledge and on the ability to perform under stressful conditions. The program emphasizes the development of good followership skills and an understanding of military customs, courtesies, and responsibilities. This is an important step in the institutionalizing process, because a clearly articulated culture must be communicated to the cadet before strong institutional bonding can occur.

Successful completion of the rigorous fourth-class system results in another rite of passage whereby cadets are "recognized" as upperclassmen. Recognition has a strong integrative effective and is another important step toward institutional bonding.

The next phase of cadet training begins at recognition and lasts until graduation. Cadets continue to grow professionally through this three-year period, but the emphasis changes from mere acquisition of relevant knowledge to application of important concepts, particularly in leadership. Numerous activities, such as soaring and parachuting, facilitate institutional identification and bonding. Some of this identification results from sharing experiences unique to the academy. As the number of these experiences increases, cadets find (sometimes to their chagrin) that they are more like other cadets than like civilian college students. This realization typically comes the first time they go home on a break (usually at Christmas of their freshman year), and grow stronger and clearer each time thereafter. The ultimate goal of this phase of training is to internalize institutional values and to develop the skills and attitudes essential to function effectively as an air force officer. This phase is replete with "inoculations"; cadets receive many opportunities to take responsibility, but they soon realize that evaluation is not based solely on what they do, but (more important) on how their subordinates perform. They learn valuable lessons about the trade-offs between time spent on individual achievement and time spent on group goals. In short, they learn that they are part of something bigger than themselves.

Academics

The academic curriculum at the academy is broken down into two main categories, core and majors courses. The core is a specified number of courses across the four academic divisions—basic sciences, humanities, engineering, and social sciences—that all cadets are required to take. From an academic perspective, the core provides cadets with a broad, general intellectual foundation for their future development as air force officers. As an institutionalizing mechanism, the shared core courses tend once again to make cadets most like other cadets, more homogeneous within themselves, and more unlike their peer groups, who tend to specialize at civilian academic institutions.

In addition to the core, cadets have the opportunity to choose one of 23 academic majors. The majors program is conceptually the same as academic majors at civilian colleges or universities and differs from its civilian counterpart in the concentration of course work in any specific discipline. Because of the large number of courses in the core, majors at the academy offer far fewer courses in their particular disciplines than a comparable civilian program.

The majors program reflects an occupational orientation and as such may impede institutional socialization. A cadet might tend to identify with his or her academic major and as, for example, primarily an electrical engineering student rather than as an officer candidate who is gaining academic expertise as an electrical engineer. Although the majors program may appear at first glance to be occupational, a close inspection suggests that this may not be the case.

We propose that an individual can be strong in both institutional and occupational orientations; our data will support this contention. If this proposition is valid, the majors program need not detract from the internalization of institutional values. In fact, one might argue that as long as core or pivotal institutional values are maintained, the presence of peripheral occupational values may actually enhance and enrich the institution through diversification. From the organization's standpoint, a committed officer corps possessing broad academic expertise would be an invaluable resource for coping with the demands of an ever-changing, complex environment.

From the standpoint of an individual cadet, the majors program fulfills several needs. First, cadets are young men and women, and as such are seeking role integration and identification. Association with a major provides an anchor for relation to peers and others outside the institution. "What is your major?" is typically the first question a college student is asked after being introduced. Answering that they are studying physics or aeronautics is certainly easier for cadets in that situation than trying to explain what studying to become an officer means. The concreteness and focus of a major can provide a feeling of stability while institutional identification continues to develop through training and education.

Returning to our immunization analogy, we find that the majors program offers several opportunities for occupational inoculations. The first is the actual teaching of the course material in the classroom, which illustrates the importance of the "blue suit" instructor. As we stated earlier, the role of the active-duty military instructor is far greater than that of teacher and educator. The military instructor is a constant reminder that something is different in the classroom. The subject matter is much the same as at a civilian university, but the difference is what Paul Harvey would call "the rest of the story." Instructors typically do two things to help bridge the gap between the classroom and the real air force. First, they show frequently how the material being studied is relevant to air force operational activities and functions. Second, instructors not uncommonly digress from the course material by telling "war stories," personal experiences relevant to the material at hand and related to the mission of the air force. In both cases, the instructor serves as a linking pin between the classroom and the air force. In the classroom, the cadets receive a dose of occupationalism as they strive to master academic concepts and receive a good grade. At the same time, the instructor helps to shape institutional values by transposing the occupationally oriented material into a broader institutional perspective, constantly reminding the cadets that the subject matter has institutional importance and that they have a responsibility to the institution to employ their new skills in the institution's best interest.

The opportunity for occupational inoculation provided by the majors programs concerns social comparison. *Social comparison* is a reality-testing mechanism whereby people engaged in particular activities evaluate the meaning and worth of their efforts by assessing the benefits received by others who are engaged in similar activities. For cadets, this inoculation occurs when cadets interact with peers at other univerisities who are pursuing similar majors. The ensuing conversations made cadets aware that they perceive the importance of the subject differently from their peers. They place an additional value on the course material because they perceive that it is relevant to their performance as an officer. Thus, social comparison distances the cadets from their peers on the outside and bonds them with other cadets, who share similar orientations.

Athletics

The athletic program has elements of both institutionalism and occupationalism. Athletes generally compete in sports because they enjoy the competition and the sense of accomplishment. They strive to improve and perform at their individual best. Participation in intramural or intercollegiate sports, however, which is required of all cadets, emphasizes teamwork as well as healthy competition. Anyone who has ever competed in sports knows that, although outstanding individual performance can be highly rewarding personally, the feeling of accomplishment is generally diminished if the team loses. The inoculation in this case is the continued exposure to situations in which outstanding individual performance is emphasized and desired, but is meaningful only if it contributes to and results in a victory for the team. A sense of the value of belonging to something greater than oneself is a cornerstone of institutionalism and makes the athletic program central to the academy socialization program.

Athletics, like the academic majors program, provide an opportunity for individual expression, identification, and role clarification. Also, because a large portion of the athletic staff are "blue suiters," athletic activities perform the same linkage function as academic classroom instruction. Finally, many people seem to associate athletics with performance in war. This idea is expressed in a quote by General Douglas MacArthur, which all cadets are required to memorize: "On the field of friendly strife are sown the seeds that on other days and other fields will bear the fruits of victory." Whatever the relationship may be between athletics and war— the importance of leadership, competition, or teamwork—a link appears to exist between athletics and the profession of arms.

MEASURING INSTITUTIONAL AND OCCUPATIONAL VALUES

What kind of values do young people bring to the Air Force Academy? How much do these values change during cadet training? Are they related to important criteria at the academy? Do men and women differ on measures of institutional and occupational values? Do different classes differ over the four years they spend at the academy? These are just a few of the questions that we tried to answer as we studied cadets' institutional and occupational values.

In November 1985, we conducted a survey of the cadet wing to measure the organizational climate in each cadet squadron, the perceived leadership characteristics of cadet leaders in the squadrons, and the perceived leadership characteristics of Air Officers Commanding (AOCs), the officers who are in charge of each squadron. In addition to these items, we asked the institutional and occupational value questions that are shown in Table 9-4. All the questions in the survey were answered on a six-point Likert scale, which ranged from "strongly disagree" to "strongly agree." The midpoint on this scale was 3.5. Four questions were combined to form an institutional scale, and one question served as our measure of occupational orientation.

Table 9-5 which shows the relationship between the items and the institutional scale, reveals two noteworthy points. First, the institutional items show an average item-total correlation with the total scale score of .61 or greater. Second, the occupational item has a zero bivariate correlation with the institutional scale. This finding indicates to us that these issues are separate; that is, an occupational orientation does not seem to be the opposite of an institutional orientation, at least in this population of cadets. If that were the case, we should see a high negative correlation with the institutional scale: as institutionalism goes up, occupationalism would go down. Because the two

TABLE 9-4. Item-Total Correlations of Items Related to Institutional-Occupational Value Orientations Among Air Force Academy Cadets

Correlation with Institutional Scale*	Statements
	Occupational
−.01 (2170) $p = .21$	In general, I look upon the air force as a place to have a good job, receive a "fair day's pay for a fair day's work," learn good skills, have a good occupation, etc.
	Institutional
.61 (2191) $p = .000$	I look upon the Air Force almost like a *vocation*; that is, it is a "calling" where I can serve my country, where duty transcends self, where obligation to country transcends particular skills I may learn or money I am paid.
.66 (2191) $p = .000$	Differences in rank should not be important after working hours. (Reverse scored)
.74 (2191) $p = .000$	I tend to identify more with civilian college students than with the officer corps. (Reverse scored)
.71 (2191) $p = .000$	Military rituals and traditions ought to be less important in today's technical armed forces. (Reverse scored)

*Institutional scale is the average score of the four institutional statements. These statements were measured on a 1–6 scale with 1 indicating "strongly disagree" and 6 indicating "strongly agree."

values are independent, a cadet apparently could be either high or low on both institutional and occupational values. This finding makes sense in light of what we have tried to point out in the first half of this chapter. Cadets have a very high need for achievement. This need manifests itself to some degree in occupational concerns because these concerns offer tangible signs of success, but not necessarily at the expense of institutionalism.

Table 9-5 illustrates several important points. First, it shows cadets' institutional and occupational orientation (as measured by the questions in Table 9-4) by class and sex. Second, it shows the overall differences by sex for both institutional and occupational orientations.

The overall average institutional scores from the freshman through senior years are 4.31, 4.17, 4.23, and 4.27 respectively. Although differences exist between the classes, no strong trend seems to be evident. Entering freshmen seem to possess roughly the same degree of institutionalism as seniors who are ready to graduate.

On the other hand, males and females differ significantly on their institutional scores during their freshman (4.28 versus 4.50) and sophomore (4.14 versus 4.35) years, but these differences decrease during their junior and senior years at the academy. The differences between the men and the women are not statistically significant during their last two years. Therefore, when we compare all men in the sample to all women (the two bottom lines of the first half of Table 9-5), we note a significant difference. We know from the previous comments that this difference is due mainly to differences during the freshman and sophomore years.

Another notable point is that the cadets in general are relatively institutionally oriented (see the top half of Table 9-5). The scores indicate that our cadets are institutional, but they certainly aren't overwhelmed by that orientation. Therefore, the Air Force Academy's insistence on trying to inculcate these values is apparently justified. Young people appear to enter the academy with moderately strong institutional values, which the academy tries to focus, increase, and strengthen.

Table 9-5 (bottom half) shows that cadets are also highly occupationally oriented. Again we see a significant difference between males and females in their response to the statement: "In general, I look upon the air force as a place to have a good job, receive a 'fair day's pay for a fair day's work,' learn good skills, have a good occupation, etc." Female freshmen and sophomores score significantly higher than the males on this question. These differences show no statistically significant differences during the junior and senior years, though the average score for women is still somewhat higher: the occupational scores across the class range from 4.09 (freshmen) to 4.29 (senior). We might suggest that cadets are becoming more occupational, but we are not sure that this is true. We feel fairly comfortable, however, in saying that cadets are both occupational and institutional in their values.

In summary, Table 9-5 indicates several important items. First, cadets score relatively high on both institutional and occupational measures. Second, over time men and women become more like each other on both institutional and occupational measures. In particular, the men's scores change to become more like the women's, particularly on the occupational measure. This merging occurs at about the junior year. At the beginning of that year, cadets must make an official commitment to the air force; they take an oath to serve as enlisted personnel in the air force if they don't finish at the academy. This oath may affect the apparent homogeneity that begins around the junior year.

Correlates of the Institutional and Occupational Measures

Table 9-6 shows the relationship between I/O scores and attitudes about significant academy experiences. Note the bivariate correlations between the occupational and institutional scores and the scores on squadron satisfaction and Air Force Academy satisfaction. Cadets who score high on satisfaction with their squadron (the approximately 125 cadets they live with during the school year) are agreeing with such statements as they are glad to be in their squadron, they are proud to be part of their squadron, and they feel good about being in their squadron. Cadets who score high on the Air Force Academy satisfaction scale are agreeing to such statements as they do not frequently think about leaving the Academy and they do not often get discouraged about

TABLE 9-5. Institutional Orientation by Class and Sex[a]

		Mean	SD[b]	N	F = Ratio	Statistical[c] Significance
FRESHMAN	TOTAL	4.31	.90	808		
	MALE	4.28	.90	711	4.89	.027
	FEMALE	4.50	.87	97		
SOPHOMORE	TOTAL	4.17	.97	634		
	MALE	4.14	.99	540	3.68	.055
	FEMALE	4.35	.88	94		
JUNIOR	TOTAL	4.23	.94	426		
	MALE	4.23	.95	369	0.11	ns
	FEMALE	4.19	.93	57		
SENIOR	TOTAL	4.27	.92	329		
	MALE	4.24	.93	282	2.85	ns
	FEMALE	4.48	.83	47		
TOTAL	MALE	4.23	.93	1915		
TOTAL	FEMALE	4.39	.87	298	7.74	.005

Occupational Orientation by Class and Sex[a]						
FRESHMAN	TOTAL	4.09	1.32	902		
	MALE	4.04	1.34	785	7.02	.008
	FEMALE	4.39	1.14	117		
SOPHOMORE	TOTAL	4.19	1.29	689		
	MALE	4.12	1.32	581	11.50	.000
	FEMALE	4.58	1.02	108		
JUNIOR	TOTAL	3.76	1.32	452		
	MALE	3.73	1.33	391	1.18	ns
	FEMALE	3.93	1.22	61		
SENIORTOTAL		4.29	1.20	344		
	MALE	4.27	1.22	292	0.48	ns
	FEMALE	4.40	1.12	52		
TOTAL	MALE	4.04	1.33	2065		
TOTAL	FEMALE	4.35	1.13	343	17.35	.000

[a]Institutional orientation is measured by combining questions shown on Table 9-4. Occupational orientation is a single item. Scale ranges from 1 to 6. Midpoint is 3.5.
[b]SD indicates standard deviation.
[c]Statistical significance or probability of occurrence.

being a cadet at the Air Force Academy. The occupational score has a very low correlation with both squadron satisfaction (.11) and Air Force Academy satisfaction (−.01), but the correlation between Air Force Academy satisfaction (.41) and institutional scores and squadron satisfaction (.23) appear to be much more substantial. This finding indicates to us that the Air Force Academy stresses institutional values. Cadets who tend to score higher on the institutional measure also tend to be more satisfied with their squadron and with the academy.

The final four lines of correlation coefficients in Table 9-6 give us pause: the correlation between the occupational measure and the rating on the value of squadron activities as valuable to leadership preparation was only .12 and the correlation between the institutional measure and squadron activities as

TABLE 9-6. Correlates of the Institutional and Occupational Measures at USAFA

VARIABLES/ITEMS	Occupational			Institutional		
	r	N	sig	r	N	sig
Squadron satisfaction	.11	2353	.000	.23	2202	.000
AFA satisfaction	−.01	2372	.238	.41	2218	.000
Squadron experience is best leadership preparation	.12	2291	.000	.28	2159	.000
Academic experience is best leadership preparation	.12	2324	.000	.02	2192	.090
Athletic experience is best leadership preparation	.12	2321	.000	.04	2188	.026
Outside activities are best leadership preparation	.10	2077	.000	.13	1968	.000

valuable to leadership was .28. These figures indicate that cadets who are institutionally oriented are more highly motivated to participate in squadron leadership activities and to see them as having some value. We were also interested to note the correlation between each of these value positions and the perceived value of the academic program as leadership preparation: cadets with an institutional perspective do not see the academic program as a valuable leadership preparation (r = .02).

We are intrigued with the implications of these findings for the day-to-day operation of the Air Force Academy and the orientation of the academic program. Cadets who are institutionally oriented perhaps do not see the applicability of transferability of their values in the classroom as readily as when they are participating in squadron activities. This possibility means that the academic program needs to turn more toward "relevance" to reach these students; otherwise, cadets with the value orientation that the academy wants will reject the cognitive tools that the academy gives them to use as officers. As will we discuss shortly, 80 percent of the cadets attain scores of 4 or higher on the institutional scale. Thus, some work is needed to make academics appear valid for leadership preparation.

Another important point is the consistently low but statistically significant correlations that are spread among the four alternatives for leadership preparation under the occupational column of Table 9-6. All the correlations are about the same, which indicates that these four choices are regarded as contributing equally well to leadership preparation. Although we can't prove it with the data at hand, we think that this pattern of scores relates directly to our hypothesis that a broad overlap exists between a cadet's need for achievement and what is perceived to be an occupational concern.

If we understand the need for achievement correctly, high-need achievers are very rational people. In their quest for challenging tasks, personal goals, and immediate feedback, they make conscious choices about where to apply their effort. The academy must offer several areas of tangible success if the students are to remain motivated. As cadets participate in each academy activity, they assess the probability that they will attain maximum success in that activity. In areas where they see a great chance of success, they participate fully. In areas where they perceive success requires more effort than is warranted by their assessment of probable success, they make the necessary effort but not the maximum effort. Their history of personal motivation demands success, and to attain success, they distribute their energies over time in those areas that give them the greatest sense of personal accomplishment. Some cadets put their efforts into the academic program, other into the squadron, and others into athletics; still others achieve their leadership success in outside activities such as clubs and choirs. By their very nature as end states, institutional values simply don't provide the immediate feedback that high-need achievers require to stay motivated for the tasks they must accomplish. Learning its values of long-term institutional rewards takes time.

Combining Institutional and Occupational Scores

Previous tables presented institutional and occupational scores by themselves, but examining groupings of cadets in various combinations of institutional and occupational scores may also be instructive. Earlier in this chapter, we argued that the occupational orientation of cadets may simply be related to their high need for achievement; because occupational matters are tangible and high-need achievers typically need specific challenges and direct feedback, cadets can be occupational as well as institutional. In fact, 81 percent of the cadets responding to the survey fell into the high-institutional group, and approximately 72 percent fell into the high-occupational groups.

If institutional and occupational scores represent two distinct sets of values, we may see a different pattern when we combine the two sets of scores and classify cadets accordingly. Table 9-7 shows cadets placed into four groups on the basis of their institutional and occupational scores. Group I consists of cadets who scored 4 or higher on both institutional and occupational measures. This is a very large group, comprising nearly 58 percent of all cadets. Group II consists of cadets who scored 4 or higher on the institutional scale but 3 or less on the occupational measures. This is the second largest group, which makes up 23 percent of the sample. If our previous data (Table 9-6) are accurate, this group might be particularly turned off by the academic program but highly motivated by squadron activities as a source of leadership preparation. Only 5 percent of the cadets fall into Group III, which scored low on both the institutional and the occupational measures. This figure suggests that the academy is successful in recruiting and selecting cadets who can achieve a good fit with the academy environment. Group IV consists of cadets who scored 4 or higher on the occupational measure and 3 or less on the institutional scale. This group makes up 14 percent of the sample. These cadets may represent high-need achievers who simply don't possess the value orientation toward the air force that the academy desires, but they may also

TABLE 9-7. Percentage of Cadets in Different Institutional/Occupational Combinations, by Class

Combination		Freshman	Sophormore	Junior	Senior	Total
GROUP I HI-I[a] HI-O	N[e]	484	376	211	203	1283
	%	59.8%	59.2%	49.4%	62.1%	57.8%
GROUP II HI-I[b] LO-O	N	195	115	135	63	515
	%	24.1%	18.1%	31.6%	19.3%	23.2%
GROUP III LO-1[c] LO-O	N	31	49	21	22	114
	%	3.3%	7.7%	4.9%	3.4%	5.1%
GROUP IV LO-I[d] HI-O	N	99	95	60	50	308
	%	10.5%	15.0%	14.1%	15.3%	13.9%

[a]Cadets scoring 4 (on six-point scale) on both occupational and institutional measure.
[b]Cadets scoring 4 on institutional measure and 3 on occupational measure.
[c]Cadets scoring 3 on both institutional and occupational measures.
[d]Cadets scoring 3 on institutional measure and 4 on occupational measure.
[e]N = the number of cadets in that group.

respond more positively to the academic program than to the military training because they were rewarded for success in this type of endeavor before they entered the academy. Therefore, the academic program would tend to meet their prior expectations as the way to succeed in life.

Overall, the data indicate that the majority of cadets have high institutional values. They also show that many cadets possess both high institutional and high occupational orientations, which may be necessary if they are to succeed in the academy's highly demanding environment.

One anomaly of the data in Table 9-7 is the pattern of percentages among cadets in their junior year. Only 49 percent of the juniors are in Group I, for example, while 31 percent belong to Group II. These data cannot be explained by the available information. The reason could be a cohort difference, something that happens during the junior year, or a function of the sample that responded to the survey.

Tables 9-8 and 9-9 indicate that the combination of institutional and occupational values has a greater impact than either value alone. We asked cadets to rate how well they thought each of four different areas (squadron, academics, athletics, and outside activities) contributed to their leadership preparation. In all cases, Group I (cadets with high institutional and high

TABLE 9-8. Differences Between Cadets, Grouped by their Occupational/Institutional Scores, Regarding which Aspect of the Academy Offers the Best Leadership Preparation

Best Leadership Preparation		Institutional/Occupational Group[a]					
		I Hi-I/ Hi O	II Hi-I/ Lo-O	III Lo-I/ Lo-O	IV Lo-I/ Hi-O	Total Group	F Sig
Squadron	X̄	3.98	3.79	2.98	3.38	3.80	38.39
	SD	1.17	1.20	1.29	1.29	1.20	.000
	N	1233	501	110	297	2141	
Academic	X̄	4.03	3.76	3.68	3.87	3.93	8.57
	SD	1.12	1.23	1.28	1.29	1.18	.000
	N	1254	509	111	398	2171	
Athletic	X̄	3.96	3.80	3.66	3.97	3.91	4.33
	SD	1.07	1.20	1.32	1.23	1.14	.000
	N	1249	510	110	299	2168	
Outside Activity	X̄	3.84	3.65	3.27	3.59	3.73	10.39
	SD	1.07	1.25	1.28	1.18	1.16	.000
	N	1126	468	102	254	1950	

[a]See Table 9-7 for a description of these groups.
Scale = 6 pt. 3.5 = midpoint. 3.5 or higher describes positive relationships. X̄ = mean.

occupational scores) rated *all* aspects of the academy higher than did the other three institutional/occupational groups. Group I cadets, for example, rated the squadron at 3.98 as contributing to their leadership preparation, while Group II rated the squadron at 3.79. Group IV rated the squadron at 3.38; Group III cadets, who were low on both institutional and occupational value orientations, rated the squadron at 2.98. The one-point difference between Group I and Group II is remarkably large.

The ratings of the academic area are worth discussing. While Group I scores the highest (4.03) of all the groups, as expected, Group IV (low institutional scores and high occupational scores) gives the second-highest rating (3.87) to the academic program as contributing to leadership preparation. This finding should not be surprising. As Table 9-6 showed, institutional scores have a zero correlation ($r = .02$) with the rating of the academic program as contributing to leadership preparation, and the occupational scores have a slightly larger ($r = .12$, $p = .000$) correlation with the academic program. The addition of a high occupational value orientation seems to contribute significantly to a cadet's perception that academic work is valuable for leadership preparation. Yet, this is a troublesome result; this group of cadets may respond positively to the academic program for the wrong reasons. They may not wish to learn how they can become better officers, but only how to be better technocrats. On the other hand, because they may have greater admiration for faculty members than for other officers at the academy, the faculty bears a greater responsibiity to use their influence with these cadets to help them understand the larger purpose of the institution.

Finally, cadets in Group III (low occupational and low institutional scores) respond less favorably to all aspects of the academy program as contributions to leadership preparation. Group III seems to consist of cadets who have no value basis for their motivation to be at the academy. Fortunately for the academy, this group is very small.

Table 9-9 shows the impact of institutional and occupational values on satisfaction and individual effort. Group I rates all three of the variables consistently higher than Group IV and Group III. Group I, for example, rates squadron satisfaction at 4.94, Group III rates it at 4.71, Group IV rates it at 4.44, and Group II rates it at 4.25. In regard to satisfaction with the Air Force Academy, Group II (high institutional and low occupational) has the higher score (4.45). Recall that institutional scores correlated at .41 with Air Force Academy satisfaction but that the occupation score was uncorrelated ($r = -.01$). This finding shows the impact of institutional values on feelings about the academy. Note also the low scores in satisfaction for Groups III and Groups IV: the scores for both these groups are negative on our scales, on which 3.5 is the theoretical neutral point. Being highly occupationally oriented is not enough; the satisfaction scores for Group IV are nearly as low as those for Group III. Evidently, institutional values are necessary for satisfaction with the Air Force Academy. Finally, Table 9-9 shows how these different groups rate their own extra effort. (This scale was taken from Bernard Bass's 1985 work on "transformational leadership.") Cadets who score high on this scale agree with statements such as "This semester I have done more than I thought I *could* do," "This semester I have done more than I thought I *would* do," and "This semester my motivation to succeed has been heightened." Again, the most common trend holds true: Group I cadets rate their extra effort highest (3.97), and Group III cadets rate their extra effort lowest (3.43).

CONCLUSION

In this chapter, we demonstrated that the Air Force Academy is an institutionally oriented organization. Our data indicate that the recruitment, selection, and training program support the institutional goals, but we believe that possessing institutional values is not enough. Our cadets seem to possess both institutional and occupational values, and the interaction of these values leads to the highest performance and greatest satisfaction at the academy. Cadets at the Air Force Academy possess a high need for achievement, and they remain motivated by what seems to be occupational rewards. Furthermore, we suggested that the academy can inoculate cadets against these occupational propensities by presenting them in a controlled environment. Thus, we hypothesize that cadets can learn when and how to subjegate occupational tendencies to institutional goals.

Further research in this area should attempt to measure need for

TABLE 9-9. Differences Between Cadets, Grouped by their Occupational/Institutional Scores, on Measures of Satisfaction and Motivation

Best Leadership Preparation		Institutional/Occupational Group[a]					
		I Hi-I/ Hi O	II Hi-I/ Lo-O	III Lo-I/ Lo-O	IV Lo-I/ Hi-O	Total Group	F Sig
Squadron	X̄	4.94	4.71	4.25	4.44	4.78	29.06
Satisfaction	SD	.98	1.11	1.22	1.24	1.06	.000
	N	1268	503	109	304	2184	
Air Force	X̄	4.35	4.45	3.20	3.28	4.17	92.46
Academy	SD	1.24	1.23	1.40	1.35	1.24	.000
Satisfaction	N	1271	512	112	303	2198	
Extra	X̄	3.97	3.91	3.43	3.66	3.89	13.66
Effort	SD	1.05	1.07	1.01	1.11	1.06	.000
	N	1247	504	108	296	2155	

[a]See Table 9-7 for a description of these groups. X̄ = mean.

achievement and to refine the occupational and institutional measures we employed. Our relatively crude measures support the contention that the Air Force Academy is a value-building and value-clarifying experience for the cadets. Yet, although the current program is highly effective, the relationships among need for achievement, occupational and institutional values, and the environment of the Air Force Academy need to be explored. Doing so may enhance the quality of an institution that dares to aspire to inculcate students with the value of duty, honor, country, and service before self rather than teaching them to expect rewards for every job well done or immediate gratification for the completion of meaningless tasks.

NOTES

1. Charles C. Moskos, "From Institution to Occupation: Trends in Military Organization," *Armed Forces & Society* 4 (1977): 41–50.
2. Gary Dessler, *Organizational Theory: Integrating Structure and Behavior*, 2nd ed. (Englewood Cliffs, N.J.: Prentice-Hall, 1987), pp. 365–660.
3. David C. McClelland, "That Urge to Achieve," in David A. Kolb, Irwin M. Rubin, and James M. McIntyre, *Organizational Psychology: A book of Readings*, 4th ed. (Englewood Cliffs, N.J.: Prentice-Hall, 1979), pp. 73–80; David C. McClelland, "Characteristics of Achievers," in Keith Davis and John W. Newstrom, *Organizational Behavior: Reading and Exercises*, 7th ed. (New York: McGraw Hill, 1985), pp. 41–44.
4. Don Hellriegel, John W. Slocum, and Richard W. Woodman, *Organizational Behavior*, 4th ed. (St. Paul, Minn.: West Publishing Co., 1986), pp. 179–183.
5. *The College Digest*, 12, no. 2, (December 1985): 7.
6. *Dash One: A Guide to Cadet Training and Development 1984–1985*. (Commandant of Cadets, Headquarters USAF Academy, Colorado Springs, Colo.).

Part Three

Comparative Perspectives

X

Great Britain

CATHY DOWNES

Considerable research has been devoted to examining the proposition that the American military is moving from an institutional format to one that increasingly resembles an occupation. If we accept that the American armed forces are, to a lesser or greater degree, marching along the path to occupationalism, we have an obvious interest in asking, "Are the armed forces of other nations following the beat of the same drummer? If they are not, what path has been chosen, and what factors have determined the choice?"

To address these questions in regard to the British armed forces, we have taken a macro approach. Such an approach analyzes the operations, policies, and internal organization of the British armed forces within the context of their relationship with the parent society. Institutional and occupational (I/O) trends within the armed forces are highlighted successfully through this focus upon the historical, political, and social interrelationships between armed forces and society.

ARMED FORCES AND SOCIETY: RELATIONS AND TRENDS

In periods of other than dire and direct threat to the integrity of the realm, the armed forces of the United Kingdom have often been regarded by various groups within the civil community as a marginal occupation, a temporary imposition, or a necessary evil. Despite these attitudes, the armed forces, particularly the officer corps, have been regarded throughout their history as an institution, an inherent and essential part of the ruling establishment and the social fabric. The contemporary military profession and its relationship with the parent society reflect a centuries-old alliance with the institutions of monarchy, Parliament, the Church, and other lesser ruling institutions. By its inclusion within this association of ruling groups, the military profession has been insulated to a considerable degree from many of the internal and external influences that have acted upon the armed forces of the United States. This curious combination of marginality and centrality is perhaps the most important characteristic of the British armed forces and their relationship with civilian society.

For most of its history, the armed forces have occupied a rather peripheral position in British society. They have taken center stage only during two world wars, when a considerable degree of integration was effected between civil and military communities At other times, the maintenance of small standing land forces was considered sufficient to counter internal security threats and to garrison a far-flung colonial empire. Moreover, the English Channel and the "wooden walls" of the Royal Navy preserved the reality of invincibility. Even when military service was regarded as a universal obligation, forces were raised only for specific emergencies. From the seventeenth century onward, military service ceased to be seen even as a temporary duty.

Civilian society and its political elite were content to leave the military profession to those who felt some sense of noblesse oblige and to others with no sense or wit at all. The rank and file of the army, and to a lesser extent that of the Royal Navy, were drawn from that group within society that had few other employment opportunities. Such men existed on society's periphery, both socially and economically. By contrast, they tended to be commanded by officers drawn from society's center, the social and usually the economic elite. A unique relationship developed between this class and military service that contributes much to the character of the profession and its institutional relationship with civilian society.

Under the feudal system of government, monarchs rewarded loyal service with ennoblement and/or grants of land. In return, landowners and lords were obliged to provide the Crown with material and manpower if the need arose. In fourteenth- and fifteenth-century Europe, the practice of organizing and leading troops on a contractual basis for a monarch or other warlord was widespread and formalized as the main method of recruiting armies. In England, the relationship between the monarch and those officers commissioned to raise regiments became permanent. Parliament regarded such a relationship as both political and financial good sense, as Sabine notes:

> Typically, colonelcies were conferred on territorial magnates who could be relied upon first, to make some material sacrifice for the sake of the prestige conferred by the colonelcy, and second, to ensure attachment of their county or region and especially its socially important elements to the Crown.[1]

The practice of purchase spread to include all commissioned ranks of the infantry and cavalry up to the rank of lieutenant colonel.[2] Purchase was seen as beneficial to the service and society on economic, political, and military grounds. Militarily, the practice ensured an officer corps drawn almost exclusively from that social stratum that was characterized not only by wealth but also by the social niceties of gentlemanly conduct. Rudyard Kipling summed up the belief neatly:

> Speaking roughly, you must employ either blackguards or gentlemen, or best of all, blackguards commanded by gentlemen, to do butcher's work with efficiency and dispatch.[3]

Economically, the purchase system made sense. If a gentleman could afford to purchase a commission, he possessed the means to clothe, feed, and equip himself for his military duties. Moreover, the Treasury did not feel obliged to provide living wages for officers, nor did it expect to pay them pensions. When an officer retired, he sold his commission, and the recouped sum served in lieu of a pension. The political argument stressed that the purchase system created a politically docile officer corps that, because of its social origins and financial investment, would be highly supportive of the status quo.

The gentlemen expressed many of the characteristics of the ruling classes and, as such, provided a vital connecting medium between the military and other ruling institutions of society. He was expected to have a private income of reasonable means. Financial independence, assisted by the interpenetration of ruling elites, ensured considerable occupational mobility; the gentleman was relatively free to engage in a range of pursuits. Enjoying the freedom to engage or not to engage in paid (however poorly) employment, and given the choice of several socially prestigious alternatives, the military officer joined his contemporaries as a leader by birth, breeding, and education. He was an amateur and proud of it.

The tradition of amateurism rested on the assumption that the gentleman, by birth and upbringing, had the requisite talents and abilities to lead and command others. Unlike the professional, the gentleman amateur did not require longer specialized training in specific skills to practice his chosen profession. The code of the amateur rested on a curious yet glorious rationale: should the amateur, depending solely on his innate talents, succeed over the trained professional, the victory must be much greater, for he had triumphed over seemingly overwhelming odds. Determination and fortitude became valued as decisive qualities in achieving the ultimate moral triumph; one might lose a battle, but never the war, and the triumph of the amateur over the professional could be equated with the triumph of good over evil, of David over Goliath.

Reforms in the army were slow in coming and were signaled by the end of the purchase system in 1871. The system was replaced by a set of competitive examinations to the Royal Military College, Sandhurst. This change reflected a hesitant and reluctant move toward a professionalism based more equitably upon both the standards of technical competence and the notions of public service and the gentleman amateur. Because the base of officer recruitment was broadened, and because of a parallel trend toward specialization in all areas of employment, army officers regarded military officership increasingly as a career, and the traditional occupational mobility of the ruling class declined.

Parliament regarded the Royal Navy as the first line of defense, and continued to see the army as a marginal institution. Therefore, they believed it should be made as "cheap to its employer as possible."[4] Faced with such hostility, indifference, and parsimony, both officers and men created a separate home and family within the regiments of the Army, often rejecting a world that had rejected or, at best, ignored them:

> For officers, the regiment was a private exclusive club, a fitting home for gentlemen. For officers and other ranks alike it was a clan, a hierarchical extended family offering a meaningful place in life.[5]

It was an army in exile by reason of its colonial duties, its location in home garrisons geographically isolated from population centers, and its own carefully maintained turtle shell, which served to make and keep it

> . . . a class of men set apart from the general mass of the community, trained to particular uses, formed to peculiar notions, governed by peculiar laws, marked by peculiar distinctions.[6]

Its only vital, impervious links with mainstream society lay in the continued status of military officership as a gentlemanly and socially prestigious pursuit and in the ties betwen the profession and other ruling institutions, particularly the monarchy.

As with the army and society, the development of the Royal Navy and its relationship with the civilian community reflects a curious mix of centrality and peripherality. Being essentially a sea force, members of the navy rarely came into prolonged contact with the civil populace they defended. By reason of its frontline national security role, however, the sweepings of soil and gutter were not considered sufficient to man the navy's ships. The nation submitted regularly to the press-gangs that conscripted civilians to man the fleets. Although the navy was out of the public sight for long periods, Britain's economic dependence upon overseas trade kept it in the public mind. A merchant ship lost to privateers or to the nation's enemies meant the difference between profit and loss to Britain's economic man, whereas a skirmish on the northwest border of India affected only the families of those killed or wounded in the action.

The ship's master, although skilled in navigation and ship propulsion, was seen to be of a lower social class than the commissioned officers, who held command by right of their gentle birth and upper-class connections. Military sea operations, however, required a technically competent officer corps, and a professional officer corps was developed under the guidance of Secretary of the Admiralty Samuel Pepys. Thus in 1678, over 150 years before the army established similar commissioning requirements, Pepys tied the award of a lieutenant's commission to the passing of an examination in which officer candidates were required to satisfy a board of senior naval officers as to their character, gentlemanly status, and seamanship. Through the establishment of the lieutenant's examination, Pepys kept the navy outside the purchase system.

Because of the Royal Navy's presence in virtually all the world's oceans throughout the eighteenth and nineteenth centuries, naval officers spent much of their time at sea on lengthy commissions and came to look upon their profession as a lifetime career. This attitude was reinforced by the early age of entry into the navy. In addition, the naval officer, by reason of his training and naval socialization, tended to be limited in his choice of alternative careers.

Fourteen years into the twentieth century, the British met with their first experience of mass warfare. For the first time, the nation as a whole was mobilized to raise, equip, and supply military forces, recruited from nontraditional sources. With sizeable numbers of sons, fathers, and husbands in uniform, the public was unwilling to tolerate incompetency, insufficiency, or inefficiency in operations, organization, or material production. Furthermore, unlike the relationship between officers and men in the peacetime armed forces, the social and economic distinctions between officers and men in the citizen armies were small because of their common civilian backgrounds and the equality imposed by trench warfare.

At the end of the war, conscription was terminated in much the same manner as one discards a wet raincoat after a storm. As quickly as the nation had improvised massive armed forces and military industries, it dismantled both. The mass involvement and sacrifices of the nation generated a wave of revulsion at the human and material cost of war. Respect for, and empathy with, the military plummeted as the citizen armies were disbanded and as the society returned hastily to its peacetime pursuits. As Barnett notes, "The brief union of army and nation after three centuries was already over."[7]

World War II proved crucial in reviving public confidence in the military profession and the profession's confidence in itself. Once again, society's concern over the armed forces increased when large numbers of men became citizen soldiers, airmen, and sailors. Again, the imperatives of mass and global warfare and national participation swept away many of the prejudices and conventions that had hamstrung professional development between the wars, and gave full rein to the national attributes of inventiveness and adaptability. Unlike the experience of World War I, however, the relationship between the armed forces and society was strengthened by the conditions that prevailed in World War II. The population mobilized more completely in support of the war; the prospect of invasion and the Luftwaffe bombings raised the nation's spirit of defiance and determination. Admiration and gratitude went out to the "few" at the Battle of Britain. The fleet of small ships at Dunkirk reawakened the spirit of the militia and the citizen's duty to participate in the defense of the nation. Moreover, the conduct of the war influenced the attitude of the civilian populace toward the armed forces. In contrast to World War I, World War II was characterized by mobility, audacious attack, and speedy, occasionally dramatic, victory or defeat in battle. The British public could applaud and reward victory and, through its national character, could turn retreats and evacuation into national triumphs.

In 1945, peace brought the realization that two nations other than Britain had emerged to hold the balance of power in world affairs. Moreover, the nation was aware that the complacent and vacillating insularity of the 1920s and 1930s had contributed much to leading Britain into a second world war. National resolution must now be maintained through the acceptance of a large defense budget and standing conscripted and regular armed forces.

National service, introduced in 1947, broadened social recruitment into the services and raised public interest in the affairs of the services. Conscription, however, was tolerated, not preferred, as a method for recruiting and maintaining armed forces. In 1957, in a major revision of defense policy and world commitments, it was announced that National Service would end in 1963.

Recruiting and retaining all-volunteer forces large enough to meet existing and future commitments proved difficult. Before World War I, the military had acted as the nation's only welfare service. Unemployment, hunger, and a brush with the law had often proved to be the greatest recruiting incentives. The Grigg Committee, set up in 1957 to examine the factors affecting recruitment in a volunteer force, noted: "Altogether we can say that for the layers from whom other ranks are derived, civilian life has become more pleasanter and more rewarding materially."[8]

Moreover, national service ensured widespread knowledge of the conditions of military service. Its emphasis upon spit and polish, pervasive regulation, and boring and invented make-work left most civilians with a perception of the military as a national repository for authoritarian values. Such a value system would become increasingly inconsonant with society in the 1960s, particularly with its youthful element, who moved to embrace a more permissive life-style.

Perhaps the most important postwar feature of the relationship between British society and its armed forces has been the growth of societal respect for the military profession and the profession's reputation for professionalism. Armed with the skills and experience of counterinsurgency warfare in Palestine, Kenya, Borneo, Malaysia, and Cyprus, the armed forces moved to carry out their policing task in Northern Ireland in 1969. On all fronts, they met a no-win situation. Sensitivities over civil control of military forces prohibited the imposition of martial law and the use of certain tactics and weapons. Exceptional levels of restraint were required in often extremely provocative situations.

Despite the union and communion of the public and the armed forces during two world wars, 16 years of conscription, and action in Northern Ireland and the Falklands War, the relationship between society and its armed forces remains one of centricity and peripherality. On the one hand, the recruitment of officers and men has expanded to include a broader social spectrum. Membership has almost ceased to be regarded as a dubious alternative to unemployment or prison for other ranks and as a gentlemen's exciting sinecure for officers. Particularly in the post-1945 period, society's respect for the military has grown apace with the many and varied demonstrations of professional competence. Moreover, the participation in military duties and training by members of the Royal Family has served as a reminder of the links between military and social leadership and has demonstrated a confidence in the armed forces as a siginificant socializing and educational institution for society's royal leaders.

On the other hand, the armed forces have retained their historical remoteness from their parent society to a notable degree, both culturally and physically. In a cultural sense, the military remains both distinct and representative. Distinctions between officers and other ranks, as perceived by groups within society, are regarded as an anachronism. Yet, sizeable portions of the population retain a certain life-style and have a philosophy that recognizes these distinctions. The central-peripheral character of the armed forces and their relationship with civilian society offer an explanation for the institutional orientation of the British military. Yet when one examines the practices, policies, and trends evident in the armed forces and their relationship with society, the British armed forces clearly do not conform entirely to an institutional model. On certain variables, the military appears to have followed a highly occupational trend. Such observations are revealed in the analysis of the institutional-occupational (I/O) thesis variables developed by Charles Moskos.

INSTITUTIONAL/OCCUPATIONAL THESIS VARIABLES

Legitimacy

The legal sanction afforded to armed forces is based upon an awareness that military service meets one of the vital needs of society. Because of these needs, and because of the consequences of the abuse of military power, society requires its armed forces to demonstrate a commitment to altruistic service: the interests of society must be placed before self-interest. The most significant sign of altruism is the subordination of individual financial rewards to the goal of serving society. In this regard, the military, as a conceptual construct, is associated more closely with a model of profession than with a model of occupation.

Although many faces of professionalism associated with military service legitimate the military as an institution, armed forces are legitimated, to varying degrees, in terms of the economic marketplace. Where armed forces are recruited voluntarily, the influence of the marketplace is unavoidable;

young men are free to seek employment in one of a number of occupational alternatives. Countries with volunteer forces must rely upon groups of people within society who are attracted to military life of their own free choice, or who can be recruited to it. In such countries, the degree to which the armed forces are legitimated by market forces can be demonstrated by the nature of the effort to recruit personnel. What factors influence young people's decisions to seek employment in the military? Is military service, for example, regarded as a chance to serve one's country, or as an opportunity for a free education or a well-paid job? How do the armed forces sell themselves to the community? Does recruiting advertising stress the uniquely military features of military service, or does it concentrate on comparability with civilian sector occupations?

In the United States, Moskos has argued that the 1970 President's Commission on an All-Volunteer Force (the Gates Commission) recommended the development of a military force highly responsive to and dependent upon civilian marketplace forces. If one examines the efforts to secure personnel for the U.S. armed forces, one can see that such forces are recruited, for the most part. The military is sold by appealing to its audience. The audience is perceived as highly liberalized, permissive, conscious of individual rights, and valuing the tertiary educational and job opportunities that provide the financial and promotion prospects needed to attain the material success so highly valued in American society. Thus, recruiting appeals have tended to deemphasize the distinctively military aspects of military service.

In the United Kingdom, the circumstances surrounding a return to an all-volunteer force were only slightly more propitious than in the United States in the early 1970s. Yet, despite the poor image of military life created by national service, withdrawal from empire, and increasing numbers of career-conscious parents seeking places for their offspring in upwardly mobile occupations, the 1957 Grigg Committee did not regard marketplace inducements as the most beneficial and effective method of recruiting volunteer forces. Instead, the rationale underpinning the committee's position was that

> . . . the answer is not to use high pay or large bounties to bribe (or try to bribe) men and women who are not keen on Service life to enter the Forces; it is to make conditions of service such that they will not deter those whom service life attracts. . . . In our view, modest increases in pay will not do anything appreciable to help recruitment, and re-engagement, whereas sweeping increases . . . would bring forward only a few more good men, though possibly a large number of "Queen's bad bargains".

The committee voiced the belief prevalent in the services that certain members of the community are naturally attracted to service life and will join because of this attraction as much as the lure of trade training opportunities or good pay. The belief has been supported by a notion that a "few more good men" are of vastly greater value than a parcel of "Queen's bad bargains." Thus, in a poor recruiting climate, shortfalls in recruiting can be rationalized by the desire not to see the quality of entrants drop below set standards. In moderate to good recruiting conditions, the services can stress the positive side of a "few good men".

Almost regardless of the variability of the recruiting climate, the services tend to stress the twin themes of contract and challenge. The notion of contract implies that the individual's commitment to the institution will be matched commensurately by the institution's commitment to the individual. The notion of challenge requires potential applicants to question their suitability for employment in an organization that is being sold as an exclusive elite, one that places distinctive demands upon its members. The challenge is presented through comparison with civilian employment opportunities. For example, an army recruiting advertisement for officers opens with the statement: "An alternative to the company car, plus office and expense account lunches." An advertisement for other ranks announces boldly: "Your feet will ache. Your body will ache. We'll even make your brain hurt."

Recruiting efforts are designed primarily to affect the thinking of certain groups within society, but there is also an element of feedback into the services. If Society believes that the person in uniform is no different than any other worker, service people naturally think of themselves in those same terms. The opposite can also hold true; the character of recruiting pitches can influence how the serviceperson views his or her employment and the relationship to the institution. In the British case, a certain satisfaction and improvement in self-image comes from the awareness that one has committed oneself to an employment that is being sold to the public on the grounds that not everyone is good enough to "join the professionals." As Henry Stanhope notes:

> A television film . . . showed three picture sequences. One was of jeeps speeding over rough ground in Cyprus, another of a patrol in Ulster with the bricks and bullets flying thick and fast, and the third showed soldiers in arctic gear in Norway, "if you are man enough to cope with this we would like to meet you" said the commentator grimly. This is the sort of gritty advertising that the Army prefers. This is not only because it reflects all that the Army likes to stand for, but also because it is more satisfying for those who are already serving.[10]

All volunteer forces are legitimated to some extent by the economic marketplace. It is the forum within which the armed forces must compete for their personnel. Such forces cannot help but be affected by marketplace factors such as levels of unemployment, salary and wage scales, and the labor demands of growth-sector industries. Even so, the directives that determine how the armed forces compete in the marketplace have a considerable impact upon the degree to which armed forces are legitimated by that marketplace or by institutional values.

Societal Regard

Military service in defense of the nation is one of a small group of activities that serve the immediate and vital needs of the community. Thus, one can reasonably expect that society should feel a certain gratitude toward those of its members who offer to meet the need for protection from external and internal threats. Society, however, tends to be erratic and capricious in extending its esteem.

A second feature that determines the nature of societal respect for the armed forces is the mystery surrounding the professional task. Delving into the depths of the human soul, cutting into the human body in order to heal it, and taking human life to protect the integrity of other human life are all extraordinary activities. They require practitioners who understand the mysteries of their tasks. Society, aware of its ignorance and vulnerability, respects that which it does not understand and applauds those who are party to the mysteries beyond its grasp.

In regard to military service, the most fundamental feature that distinguishes the soldier from the civilian is the soldier's commitment to selfless service. Although such a commitment is also made by other groups within society, only military service involves an unlimited liability, which may result in the sacrifice of life itself. Traditionally, society has recognized and respected the soldier for his special commitment to place himself willingly in life-threatening situations on society's behalf. This respect includes the recognition that because such men make this commitment, the rest of society can life in comparative peace. Thus, peace belongs to the civilian, not the soldier.

To an increasing extent, the armed forces of the United States have received a form of societal prestige similar to that afforded to civilian occupations, rather than a prestige based upon the institutional characteristics traditionally associated with military professionalism. This development is not surprising in light of the predominant value structure of American society and the actions that have caused considerable civilianization of the armed forces.

Within American society, perhaps the most highly espoused value is a commitment to individual attainment and visual display of material success. The supporting value structure accords high priority to the notions of egalitarianism, egocentrism, and the rights of the individual to compete for success in a free-enterprise economic system, which rewards the fittest. As a consequence of these values, American society tends to accord esteem to those occupations in which participants can secure the highest financial rewards, often with no regard for the character of the occupation. The medical and legal professions, for example, ostensibly enjoy a high level of social esteem on the grounds that both professions seek to aid individuals on a basis of universalistic and altruistic service. In American society, however, how much genuine social prestige and esteem is accorded to the impoverished lawyer, the impecunious physician? In a social value system in which money and power are the currency of success and success the professed deity, these individuals are rarely seen as successful and therefore worthy of social esteem, despite the nature of their work and their commitment to serve society. Such a value system is almost completely at odds with the values that have traditionally been regarded by military forces as operational imperatives, such as a readiness to sacrifice life in the service of a common cause.

The process of civilianization acts in parts to destroy the essential mystique of the military profession. The status accorded the armed forces on the institutional criteria of patriotism, fidelity, selfless service, and dedication depends upon the armed forces being viewed by society not as a mirror image of itself, but as a distinct and distinctive entity, unassailable in its exclusive qualities and privileges. Certainly, an armed force that has become too exclusive may have a potentially lethal chemistry. An armed force that sees itself as the last moral and ethical bastion in a societal sea of greed, egocentrism, and moral and political corruption is, in some circumstances, a military coup waiting to happen. Equally to be feared, however, is a military at the other end of the spectrum that cannot make the distinction between itself and the society it serves. If no distinction is perceived, either by society or by the armed forces, then on what basis should servicepeople make a greater commitment to societal security than any other citizen group? A trend in the opposite direction, however, can create alienation, ignorance, and a military force similarly lacking in high levels of societal support. In liberal democracies, maintaining a civilian-military relationship that falls between these two extremes is a difficult task. A physical and psychological separation of armed forces and society that allows the military to retain its aura of mystery and invisibility must be created and maintained. This goal, however, conflicts with the need to ensure the military's close affinity with the predominant values and social structure of the greater society and to develop society's awareness of its armed forces.

In the United Kingdom, four factors have influenced the fragile character of this civilian-military relationship.

First, the British armed forces have historically been separated from society in economic, geographic, and social terms. The trend in this century has been to reduce but not to eliminate the gap between the two communities. Although troops are no longer stationed in large numbers in colonial outposts for example, British military bases are often located away from concentrations of civilian population. Moreover, as a community, the armed forces have both consciously and unconsciously sought to retain a distance between themselves and civilian society.

Despite the desire for privacy, the armed forces, deprived of their role as colonial police and returned to home waters and continental locations, remain sensitive to the accusation that they fulfill no productive function. As a result of this sensitivity, the services have engaged in efforts to make society more aware of them. The aim has not been to make the military recognizable in civilian terms, but to promote and communicate an image of the armed forces that stresses its military functions and its professionalism. This effort to keep the armed forces in the public eye ("KAPE") includes military skills demonstrations, involvement in community projects, and the Royal Navy (RN) and Royal Air Force (RAF) Presentation Teams.[11] The message has been reinforced in recent TV documentaries such as "Fighter Pilot" (on the selection and training of an RAF fast-jet pilot), "Sailing" (on the last cruise of the aircraft carrier HMS *Ark Royal*), and "The Paras" (on the training of new entrants to the Parachute Regiment). Each series has graphically shown the military as a distinctive community with unique skills, not easily acquired, yet composed of a group of very personable, human young people.

In contrast to the United States, Britain experiences much less incongruence between the prevailing civilian social structure and supporting value system and the structure and value system of the armed forces. In the United States, the inherent antagonism has been ameliorated and diffused by the efforts of the military to substitute civilian norms and values for military ones. In the United Kingdom, the armed forces have not been pressured to make a similar level of accommodation, mainly because of the coincidence of core military and civic values.

One can argue that these traditional civic values evolved from the impact of the historical relationship between military leaders and followers upon the development of a social class system. Considerable social change, both in society and in the military, has diluted the intensity, pervasiveness, and rigidity of these class relationships and their supporting value system. In the contemporary setting, the military continues to place a higher premium upon certain values than does society generally, but the armed forces tend basically to be respected for their adherence to these values and for the character of these values.

The underlying consonance between the military and society on value priorities and ethical standards is supported and sustained by the legitimation of the armed forces as an institution of the Crown. On a constitutional and symbolic level, the relationship with the Crown removes the armed forces from the petty functionalism of party politics and provides a distinctive legal status for the services. Her Majesty the Queen remains the titular head of the armed forces; all warrants and commissions are issued in her name. As Sabine observes:

> The armed forces are pre-eminently Crown rather than state institutions. The strong personal connection between the services and the reigning monarch . . . is sedulously maintained, and all volunteers—officers and men—are attested members. A whole panoply of ritual and symbolism is employed to sustain the importance of the personal oath of allegiance. . . . All principal ordinances are Queen's Regulations. None of this is mere anachronism.[12]

The association serves to enhance the respect and prestige accorded to the armed forces by society. Moreover, the connection stresses the commitment, however symbolic, of the "important" elements of society to the legitimacy of armed forces. An American commentator notes:

> . . . in the War of the Falklands, Prince Andrew flew a helicopter on entirely real operational

missions. By contrast, think how highly unlikely it would be to find Ronald Reagan's son or Walter Mondale's in a combat zone or even in the military.[13]

Along with the visibility of the armed forces in ceremonial duties, the influence of mass media has extended into operational activities in the 1970s and 1980s. The British armed forces have been called into action on numerous occasions since the end of World War II. These actions have provided the British viewing public with proof of service and have enhanced the military profession's reputation and prestige.

Keeping the peace in Northern Ireland, Cyprus, or the Sinai, maintaining essential services in times of disruption, fighting North Sea cod wars, relieving hostage sieges, assembling and disarming guerrillas in Rhodesia, and assisting with famine relief in Ethiopia are examples of service as a constabulary force. Such examples are often characterized by dirty and difficult fighting, the need for great restraint and forbearance, and sometimes only that very slight satisfaction of having stopped a situation from deteriorating further. With a history of such actions throughout the period of colonial garrisoning and withdrawal, one can argue that the British public, although just as intolerant of unmitigated defeat, is more aware than its American counterpart of the extreme odds against dramatic success in these encounters.

Basis, Mode, and Level of Compensation

The manner in which British servicepeople are remunerated is a curious combination of institutional and occupational aspects. On the one hand, the process for establishing wage and salary scales is responsive to adjustments in civilian marketplace conditions. On the other, the levels and nature of compensation are supported by a paternalistic rationale that recognizes that because servicemen make certain sacrifices inherent in the very nature of the job, their interests should be looked after.

This paternalistic attitude underpinned the recommendations of the Grigg Committee. Although rejecting financial inducements as the basis for recruiting an all-volunteer force, the committee recognized the necessity of assuring military personnel that their financial needs would be met. To this end, the Grigg Committee recommended the periodic scrutiny of the armed forces' pay to prevent erosion of comparability with the civilian sector.

In 1969 the National Board of Prices and Incomes recommended the creation of the "military salary" and the establishment of an independent committee, the Review Body on Armed Forces Pay (hereafter the Review Body), to review all aspects of the military salary, including pay increases, pay differentials, allowances, pay bands, and charges for food and accommodation.

The military salary has both institutional and occupational characteristics.

As a basis for compensation, it is founded upon the institutionally oriented principles of rank and seniority within rank. Yet, in keeping with the Grigg Committee requirement that servicepeople receive a "fair reward for the jobs they do," pay scales are based upon three factors. First, military pay is compared with the pay of jobs of equivalent weight in the civilian sector. Second, pay rates include an increment—the X factor—that recognizes the special features of military life that "make it more uncertain and on occasions more hazardous than the normal run of employment in civilian life."[14] Finally, the pay of the individual serviceperson is determined, in part, by eligibility for specific forms of additional pay and allowances.

Pay comparability with jobs in the civilian sector is assessed through analyses of the constituent parts of each military job by using criteria common to all tasks, such as level of responsibility, experience and skill levels management demands, and educational qualifications. With the cooperation of civilian firms, civilian jobs are similarly analyzed. For example, the 1983 Review Body noted that job evaluations had been carried out at the corporal's level for almost 600 jobs in 200 organizations and at the warrant officer's level for almost 400 jobs in 130 organizations; at the officer's level, some 400 firms were involved.[15] All aspects of pay are considered including fringe benefits and pensions. Thus the 1983 Review Board reported:

> With the exception of the benefit provided by the private use of a company car we have concluded that, on the whole, non-pay benefits available to members of the armed forces roughly balance those available to their civilian counter-parts.[16]

The average of the range of pay in each civilian-sector job is used as a guide to establish the appropriate pay level in the armed forces. Each service job is classified on the basis of size, level of responsibility, and physical and mental skill requirements, and grouped into one of seven pay bands. Because of the large variation in jobs undertaken by military-personnel within the same rank, the job-specific pay band determines the particular financial reward. Thus, pay for service people is not determined entirely by rank.

The individual remuneration of military-personnel is also determined by the type of engagement. The army and the RAF have three scales of committal pay: Scale A for those committed to serve less than six years, Scale B for those committed to serve for six or more years but less than nine years, and Scale C for those committed to serve for nine or more years. Scale B personnel under the rank of sergeant, for example, receive $205 more per year than those in Scale A, and Scale C personnel under the rank of sergeant receive $345 more per year than those in Scale A. In the Royal Navy, the principles of "open engagements" and "bonuses" operate. All ratings join on a 22-year engagement with a minimum service commitment of four years from the age of 18, and an option of giving 18 months' notice to leave at any time after the first $2\frac{1}{2}$ years of service points. Thus, they are committed to serve for six or nine years, and they receive a $1,250 lump-sum bonus at each service point. In this way, reward is given for services rendered, as opposed to payment for a commitment to serve.

Pay banding and job evaluations seek to ensure a broad comparability with civilian-sector occupations, but the Grigg Committee recognized that certain unique circumstances within military life place the military at a disadvantage to their civilian counterparts. These circumstances were defined by the National Board for Incomes and Prices as follows:

(i) The Serviceman is wholly committed to the Service and is subject to a code of discipline which reaches far beyond that obtaining in any form of civilian employment.

(ii) He is liable to be exposed to danger on active service.

(iii) He is required as part of his normal peace time service to endure bad or uncomfortable conditions while in the field or on board ship.

(vi) He is subject to the constant upheaval and uncertainty—or what has been called "turbulence"—imposed by the need for high mobility in a military force.[17]

The board recognized that these features affected all service personnel and therefore could not effectively be covered by a system of allowances. All are subject to military discipline, are liable for duty 24 hours a day and seven days a week, and are unable to resign at will, change jobs, or negotiate for changes in pay and conditions. Although only a percentage of armed forces personnel become involved in life-threatening situations, the board noted that in choosing a military career, ". . . the serviceman consciously accepts the liability to take part in military operations at some time or times during his military career."[18] As a result of the board's assessment of these factors, it was decided that a special compensation payment, the X factor, should be incorporated into the military salary. The X factor is expressed as a percentage of the basic military salary and is currently set at $10\frac{1}{2}$ percent for men, $7\frac{1}{2}$ percent for women, 5 percent for male reservists and $3\frac{1}{2}$ percent for female reservists.

Although job evaluations, pay bands, committal pay, and the X factor determine the basic level of remuneration for all military personnel, individuals receive further recompense in the form of "additional pay." Major entitlements, which are paid continuously to those serving or liable for service in particular duties, are regarded as an "incentive to recruitment and retention in certain areas of employment in the armed forces where the military salary alone does not provide sufficient inducement."[19] These entitlements include flight pay, submarine pay, parachute pay, hydrographic pay, Gurkha service pay, and Northern Ireland and Falkland Islands pay. Minor supplements, which take into account particularly difficult or trying conditions that occur on a noncontinuous basis, include payment for objectionable work, payment for work in unpleasant conditions, experimental pay, and Gurkha language pay.

Two other payments contribute to the military salary. First, married servicepeople receive a separation allowance when they are serving over 200 miles away from their homes. Second, those completing more than nine years of service are eligible for length-of-service increments depending on the years of service.

Before 1970, military personnel received free food and accommodation. Those who were not on the ration strength of their units because they were on leave or were married and living out of barracks were paid a ration allowance. Married men received a marriage allowance for accommodation, but had to pay a nominal charge for public married quarters or rented private accommodations. Under the present military salary system, all servicepeople are charged for food and accommodation.

On the surface, the military salary is institutionally based because of its determination by rank, its continued calculation as a daily rate, its paternalistic protection factor (the X factor), and its retention factors of additional forms of pay and committal pay. The salary, however, is also made up of a number of features closely associated with an occupational ethic. Wage scales are based upon job-specific pay bands within rank; the scale of pay for each pay band is determined by job evaluation of criteria common to armed forces tasks and to civilian jobs of comparable weighting. Military personnel receive extra recompense for working in unpleasant or difficult conditions in the same manner as a civilian would receive a bonus. Moreover, major forms of additional pay are directly comparable with competitive salary bidding between employers in the civilian sector to secure and retain sought-after workers. Finally, the military salary has ensured that military personnel receive a cash salary from which they are expected to meet the cost of living, as opposed to the more traditional pay arrangements in which a significant element of salary was received in noncash form.

Military Spouses and Families

Loyal doormat or demanding interloper, the military spouse has come to play a prominent part in military life as more and more servicepeople marry. In the past, armed forces tended to be composed predominantly of single men, but in the contemporary British armed forces, the traditional camp followers have been replaced by a substantial "petticoat brigade." Across all services, on average in 1984, 74 percent of British officers and 53 percent of British other ranks were married.

In conformity with its predominantly institutional character, we can fairly say that the British armed forces seek to be paternalistic employers. They exercise a benevolent, if somewhat arbitrary and occasionally insensitive, influence over all those who come within their orbit. The forces offer more than a job to their employees and provide them with more than employment. This firm employs some 326,000 people directly and 203,300 more part-time.

It maintains some 132,000 houses and rental units for its employees, 47,500 of which are located outside the United Kingdom. It educates 29,220 children of its employees in 114 schools provided by the company overseas. A further 21,700 employees' children receive assistance from the firm to attending boarding schools. The firm operates its own TV, film, and radio network, with some 25 cinemas, an estimated TV target audience of 165,000, and target radio audiences of 208,000. In addition, the firm operates 81 trading outlets and 70 video libraries for its employees.[20]

In addition to providing access to the practical amenities of life, the firm supplies an array of official and quasi-official social services and allocates personnel to assist employees and their families. On an RAF station, for example, such personnel include a families officer and staff who are responsible for the administration of married quarters. The RAF station also has a personnel services flight, a supply squadron, education and medical officers, a station warrant officer, and chaplains. In addition, the firm, in association with employees and their families, administers the many facilities that make the military community self-sufficient, such as nursery schools, libraries, community centers, sports facilities, hobbies and handicrafts clubs, and evening classes. Finally, the military community is supported directly by organizations such as the "Soldiers' Sailors' and Airmen's Families Association" (SSAFA), which aims at helping service and ex-service families in difficulty, and the Navy, Army and Air Force Institute (NAAFI), which provides shops and clubs for the armed forces in areas and conditions in which no civilian commercial organization would operate.[21]

Although the firm seeks to create and maintain self-sufficient communities of employees and their families, these communities are often in a state of flux. Between 1975 and 1980, for example, Royal Navy and Royal Marines (RM) officers averaged 2.5 moves, army officers averaged 4.0 moves, and RAF officers, 3.0 moves. During that period, RN and RM ratings moved on average 2.1 times, soldiers, 3.0 times, and airmen, 2.7 times.[22] These figures do not include shorter periods of separation, such as exercises, courses, and short, unaccompanied tours.

Aware of the disruptive effect of separation and turbulence on family life, the services have sought to minimize their effects. Two efforts are worthy of note. First, service children in the United Kingdom may attend their nearest Local Education Authority (LEA) school. On overseas postings, the services operate the Service Children's Education Authority (SCEA), which attempts to provide education facilities similar to those provided by LEAs. As an alternative, all servicemen are eligible for assistance under the Boarding School Education Allowance (BSEA) scheme, which aims at meeting most, if not all, of the fees charged for a child to attend a boarding school or a fee-paying day school while he or she is living with a guardian. Second, the services seek to enhance wives' and families' understanding of their husbands' and fathers' employment. Wherever possible, the Royal Navy arranges for families to visit ships on which the husband is serving, and various army units arrange informally for the wives to spend a few days in the field with their husbands' units while "hubby" minds the children.[23]

The high incidence of marriage in the armed forces does not in itself indicate an occupational military, nor is a military composed predominantly of single soldiers necessarily institutional. However, the institutional or occupational orientation of an armed force is revealed in its policies regarding the married soldier, the military spouse, and the military family. One would anticipate that an institutional military would retain traditional views as to the roles of women and would seek to draw wives under its paternalistic influence to ensure closer identification and affiliation with the needs of the institution. In an occupational military, one would expect to see more liberal attitudes towards the roles and employment of women and less physical and psychological control of the military spouse by the military establishment. The wife would be no more and no less involved in her husband's career and employment community than any other wife with a husband in any other occupation.

Residence

Another facet of the trend toward occupationalism in armed forces is seen in the disintegration of the institutional military community, with the separation of residences and workplaces. The occupational military would have a diminished influence over the military family that lives in a civilian community. As a homeowner, the service member joins his civilian neighbors in their concerns over house prices, taxes, landscape gardening, and neighbourhood watch groups. He joins them in their daily commute to and from their places of employment. His children join their civilian peers in attending local schools, and the family joins in the social interaction of the community of which they are a part.

Establishment of the military family within civilian communities is facilitated by a number of factors. First, military families cannot be integrated into a civilian community if one does not exist. Large military bases usually attract a range of service industries and their employees and families, and the resultant civilian community provides fertile ground for the assimilation of military familes. Moreover, the military ceases to be the sole provider of social services and amenities. Second, if the armed forces do not provide sufficient on-base housing for their married personnel, the service members have no alternative but to seek civilian housing. Third, military salaries need to be maintained at a level that supports house buying. Finally, if posting policies are based upon movement only when promotion and vacancies dictate, and upon features such as home-basing, the military family will attain the stability needed to establish roots within a civilian community.

The integrity of the military community has been premised to decline as more service families move out and become integrated into civilian communities. In the British armed forces, house ownership has increased. In 1973, 14.9 percent of army married personnel, 27.7 percent of RAF married personnel, and 49.9 percent of naval married personnel owned their own houses. By 1983, these percentages had increased to 24.0 percent, 42.9 percent, 65.5 percent, respectively.

The increased incidence of house ownership has affected the institutional character of the armed forces, but a number of factors have inhibited and countered the disintegratory effects of house purchase upon the military community. First, the impact of house purchasing upon the integrity of the military community is determined in part by the location of that community. In overseas areas and remote locations in the United Kingdom, military families tend to be more dependent upon the military for the essentials of life and more dependent upon each other for the quality of life. In the United Kingdom, for example, service families as well as civilian families come under the cover of the National Health Service. For military families overseas, however, the services take over responsibility for medical and dental care, including maternity and child welfare clinics.

Second, although increasing numbers of service personnel are buying their own homes, only a percentage of the owners actually live in their houses, particularly in the army. In 1983, for example, although 24.0 percent of married personnel owned their own houses, only half of them lived in those houses.

Third, in an occupational military, service members and their families are absorbed through house buying into civilian communities. The military family is civilianized and the military community is dismembered as families leave its physical and psychological orbit. Even so, the disintegratory effects on the military community and the changes in values and priorities of the military family caused by house buying are offset if significant numbers of service members buy their own houses in the same area. In such cases, the military family is not absorbed into a civilian community, but effectively remains within the influence of the military community.

Two factors have influenced the creation of these de facto military married patches. First, as servicemen move into the age and service groups in which they consider buying their own homes, they tend to seek houses located near particular military bases and in a certain price range. Most housing estates and building developments include houses of around the same purchase price. Therefore, service personnel can end up living next door to each other when they buy houses. Second, there are programs whereby service members can apply to purchase discounted surplus married quarters, as well as purchasing schemes for military personnel in the last few years of their service.

The variations between services in house residence and ownership reflect differences in posting locations and policies. For example, the RN has only a small number of posting locations. Under a port-basing system, the service members can assess the number of times they are likely to be based out of a particular port. On the basis of that assessment, they can seek to purchase houses located near the port to which they are likely to be posted most often. Thus, in 1980, 63 percent of all married RN/RM house owners had their houses located in their present duty area. Because of the larger number of posting locations for army personnel, only 32.3 percent of all married army house owners' homes were located in their current duty areas.[24]

Although the RN has fewer posting locations, it posts its personnel individually to billets in the duty areas; so, too, with RAF personnel and members of the technical corps of the army. In the line regiments, however, posting is often conducted by battalion. A UK-based home defense battalion with a dependent population, for example, will deploy for a tour in Germany, followed by a move to Northern Ireland. Thus, the units and families rotate together regularly. As a result, although army regimental families tend to move more often, they also move in groups, which tend to be more self-sufficient than an individual family.

Regarding the frequency of posting, the RN and the RAF follow policies that aim at minimum turbulence. In the RAF, for example, UK–based tours are mainly for unspecified lengths. After minimum time periods in a station post, servicepeople are given the option to extend their tours of duty. Thus, policies are reflected in the frequency of moves; as noted earlier, army personnel tend to move more often than their RAF and RN/RM counterparts.

The different attitudes toward house buying among personnel in each service can be seen in the responses of service personnel to the 1983 Armed Forces Accommodation and Family Education Survey question on main reasons for house purchase. Although 62.4 percent of RN/RM personnel in the survey listed "settled home" as the main reason, only 29 percent of army personnel listed this as the main reason. Similarly, only 29 percent of RN/RM officers and ratings listed "retirement" or "saving" as the main reason, whereas 62 percent of army personnel and 57 percent of RAF personnel noted "saving" or "retirement" as the main reason.[25]

The increased incidence of house purchase is perhaps the most occupationally oriented trend evident in the British armed forces. In conjunction with the effects of the increased incidence of marriage, house purchasing has altered the character of the military community. Officers' messes very often are demanded of live-in members; married service personnel can go home to their families for lunch rather than eat together. Although the armed forces have not internalized a 9-to-5-and-home-for-dinner ethic, the Monday-to-Friday-and-home-for-the-weekend practice is a consequence of military families settling in their own homes. Despite recognizable changes, however, the integrity of the military community has not been destroyed. In this instance, a seemingly occupational trend has had certain institutional consequences, such as the development of de facto military married patches.

CONCLUSION

What circumstances and factors have combined to propel the U.S. armed forces toward occupationalism and civilianization? Why should armed forces, by definition organized, legitimated, armed, and prepared to apply force and coercion in support of national goals, internalize ethics and practices that appear to run counter to the operational imperatives of that function? Are these circumstances and features endemic to the American nation, culture, and military? Can the resultant trends be seen in the armed forces of other nations? If so, do such trends hold the same potential for self-destruction that certain American commentators see in the current personnel policies and practices of the U.S. armed forces? If the experience of other nations differs from the United States, what factors have contributed to this difference? Finally, is any of this experience valuable and applicable to the American case? These questions have been raised within the framework provided by the I/O thesis.

Within the context of the British armed forces, it has proved both possible and practical to examine the current military personnel policies and civilian-military relations through the application of the I/O model. Not only does such an exercise reveal much about the character of the British armed forces and their relationship with their parent society, but also it highlights certain observations on the I/O thesis itself.

First, the institutional or occupational character of an armed force is as much a product of historical experience, cultural heritage, and civilian/ military interaction as of any specific military policy initiatives and organizational arrangements. Armed forces are the dependent variable in the relationship between the military institution and society. Referring only to factors and actions within the conscious control of those organizations is sufficient to explain changes within military organizations. It is not sufficient, for example, to note that in an institutional military, the military spouse will be more committed to and more a part of the military community because an institutional military is capable of exerting more influence over the military spouse than is an occupationally oriented military. Such a statement fails to take account of factors such as the influence of societal changes on sex roles and the ages and social backgrounds of those seeking and becoming military spouses.

In the examination of the British and American cases, the institutional military, with its historically inspired separation from its parent society, appears more capable of determining independently its own responses to societal changes rather than being dictated to by those changes. This may be another argument for the maintenance of a minimum level of separation between armed forces and society.

Second, the comparative uniqueness of national historical experiences and cultural heritage must qualify the transferability of policies and organizational arrangements from one nation's armed forces to another. The adoption of a British-style combat arms regimental system by the U.S. army is a case in point. The characteristics of the British regimental system suit the type of fighting in which the British army has most often been involved, the temperament of the British soldier, and the awareness of the psychological needs of the soldier in battle. The assumptions that make the system work, however, have suffered in the Atlantic transfer and either have failed to be incorporated or have been misinterpreted.

Third, a trend that is identified as occupationally or institutionally oriented in one nation's armed forces does not necessarily have the same connotations in the armed forces of another nation. Various factors may intervene to reverse the effects of a particular trend. The trend toward increased house ownership among U.S. service personnel, for example, has led in many instances to a separation of home and work locales and the integration of service families into civilian communities. Both conditions have reduced the distinctions between the military career and other occupations and between civilian and military communities. The same trend among British service personnel, however, has not had the same effect. The mobility of the force ensures that a proportion of house owners do not live in civilian houses, although they own them. The development of de facto married patches in civilian housing areas ensures that the military community is merely transplanted rather than disintegrated. Recognizing trends in one nation's armed forces that have been identified in the armed forces of another nation is not sufficient. Such a process needs to be supported by an analysis of the conditions under which policies are formulated and the effects of such policies and practices.

Finally, the British armed forces appear to be an example of a military that in most respects is firmly anchored in the institutional camp. Its institutional legitimacy is based strongly upon the historical alliance between the military and other ruling institutions in society. The armed forces are recognized by that society as a distinctive, unique, serving, and comparatively mysterious organization. The military profession makes demands upon its members that other forms of employment do not: extended tours of duty overseas; fixed-term contracts of employment; liability for 24-hour, seven-day-a-week service; frequent movement of service members and families; subjection to military discipline and law; lack of the prerogative to resign, strike, or negotiate over salary and working conditions; and the possibility of being involved in life-threatening situations.

As an employer, the British armed forces have proved to be institutional in their paternalistic desire to provide for the needs of military personnel and their families and to control benevolently the environment in which they and their families live, work, and grow. Within this institutional framework, occupational features have developed and can be identified, in some cases independently of the armed forces' efforts and in some cases as a direct consequence of military policy. Such features include an increased incidence of marriage and home ownership among service members, increasing numbers of military spouses seeking and securing independent employment, and the job comparability assessment aspects of the military salary. Despite the existence of such features, however, we cannot premise with any degree of confidence that these characteristics represent a trend toward occupationalism in the British armed forces. Rather, they represent areas in which the armed forces have willingly or grudgingly accommodated broad social changes in their parent society and their relationship with that society.

NOTES

1. John Sabine, "Civil-Military Relations," in John Baylis, ed., *British Defence Policy in a Changing World* (London: Croom Helm, 1977), p. 244.
2. Commissions were not purchased in the artillery or engineers. After 1741, all officer candidates for these two forces received their commissions on completion of a course of instruction at the Royal Military Academy, Woolwich.
3. Byron Farwell, *For Queen and Country* (London: Allen Lane, 1978), p. 70
4. Sabine, "Civil-Military Relations." P. 274.
5. Farwell, *Queen and Country*, 1978, p. 25.
6. William Windham, in Colonel R. D. Heinl, Jr., ed., *Dictionary of Military and Naval Quotations* (Annapolis, Md.: US Naval Institute Press, 1966), p. 14.
7. Corelli Barnet, *Britain and Her Army 1509-1970: A Social and Political Portrait* (London: Allen and Unwin, 1970), p. 410.
8. Ministry of Defence, *Report of the Advisory Committee on Recruiting* (London: HMSO, October 1958, Command Paper Number 545, Grigg Committee Report), p. 6.
9. The Grigg Committee Report, p. 18.
10. Henry Stanhope, *The Soldiers: An Anatomy of the British Army* (London: Hamish Hamilton, 1979), p. 48.
11. The RN and RAF Presentation Teams usually consist of three or four officers and support technicians. The teams tour the country giving general briefings to interested groups —community organizations, university groups, and the like—on the roles, equipment, and personnel of their respective services. In 1982–83, the RN Presentation Team recorded an invited audience totalling 37,000 people. Over eight months in 1983–84, the team gave presentations at some 61 different venues. *Royal Navy Broadsheet 83* (London: HMS, 1983).
12. Sabine, "Civil-Military Relations", p. 245.
13. Fred Reed, "Alienated Army is Our Creation." *Washington Times*, May 8, 1984, p. 3.
14. National Board for Prices and Incomes, *Standing Reference on the Pay of the Armed Forces* (London: HMSO, Command Paper Number 4079, June 1969, Second Report), p. 21.
15. Review Body on Armed Forces Pay, *Twelfth Report* (London: HMSO, Command Paper Number 8880, May 1983), p. 4.
16. Ibid., p. 5.
17. National Board for Prices and Incomes, p. 21.
18. Ibid., p. 22.
19. Ibid., p. 22.
20. Secretary of State for Defence, *Statement on the Defence Estimates 1985* (London: HMSO, Command Paper No. 9430-II), pp. 22, 24, 36, 50, 54—55. 1980 figures were used for the number of Boarding and Day School Education Allowance recipients. See Ministry of Defence (Statistics), *Report of the Armed Forces Accommodation and Family Education Survey* (1980), pp. 256 and 260.
21. The SSAFA was formed in 1985 and aims to help service and ex-service families to obtain all the assistance to which they are entitled: social security, pensions, financial assistance to those in distress, and representation of family cases to appropriate government departments. The SSAFA also operates nursing and social work services on overseas bases. The NAAFAI offers hire purchase, credit, and insurance facilities for service personnel and their families and provides bar and catering facilities in social and junior ranks' clubs.

22. Ministry of Defence (Statistics), *Report of the Armed Forces Accommodation and Family Education Survey* (1980), Table 8.3, p. 208.

23. One such outing was organized by the 4th Royal Tank Regiment and entitled "Exercise Tired Lady": 18 wives of A Squadron Soldiers spent two days in the field driving Chieftain tanks of the Demonstration Squadron on Salisbury Plain. "Exercise Trouble and Strife," organised by C Squadron, the Royal Scots Dragoon Guards, involved wives in grenade throwing, assault courses, and a death slide at the Sennelager training area. See Graham Smith, "Exercise Tired Lady," *Soldier Magazine* 40, (July 10, 1984): 22–23.

24. *Report of the Armed Forces Accommodation and Family Education Survey* (1980), p. 208.

25. Ministry of Defence (Statistics), *Report of the Armed Forces Accommodation and Family Education Survey* (1983), p. 85.

XI

Federal Republic of Germany

BERNHARD FLECKENSTEIN

In Germany, discussion of the military vocation is especially problematic and emotion-laden because of the country's history, as will be shown later in this chapter. Interest in the topic is concentrated not so much on the development of the military vocation as on the effects of the military's professional self-image on the relationship between the military and civilian society.

Charles Moskos's institution-occupation (I/O) thesis is well known in Germany and has been reviewed widely in German military-sociological literature, as have the concepts of Abrams, Feld, Finer, Huntington, Larson, Janowitz and van Doorn.[1] The paper on the I/O model that Moskos presented in 1981 at the Fourth Annual Conference of the International Society of Political Psychology has been translated into German and published.[2] Until now, however, no empirical research based specifically on this approach has been conducted in Germany. So far, the I/O thesis has been applied merely to interpret data collected for other purposes.[3]

For some time now, the debate in the Federal Republic of Germany on the nature of the military vocation has been dominated by two major approaches: the civilian-oriented approach, which analyzes the military vocation as an element of a large bureaucratic-technical organization and ascribes to it civilian occupational concepts; and the military-oriented approach, which differentiates between the functions of military and civilian occupations, thus using the specific characteristics of the military vocation as guiding principles for analysis. The advocates of the civilian-oriented approach view the soldier as an employee like any other, whereas those favoring the military-oriented position emphasize the unique features of a soldier's profession.

The confrontation between the two contradictory positions has not been emotionless. From the very beginning, the public debate on the issue was politicized; the advocates of the civilian-oriented vocational model were labeled *modern* and the advocates of the professional military model were referred to as *conservative*.

177

The debate on the proper model for the military vocation is understandable when we consider the German historical experience with the military. The conceptualization of such a vocation is linked to the soldiers' perception of themselves in society: do soldiers claim a special role for themselves as they have in the past, or are they willing to enter into a close association with the civilian population, as Scharnhorst called for 180 years ago?

REFORM AND REACTION: THE SEARCH FOR A VOCATIONAL MODEL

On May 9, 1955, the Federal Republic of Germany entered NATO, thus committing itself to contributing to the common defense of the alliance. This date also marked the temporary end of an intense five-year domestic debate concerning Germany's rearmament. Once the decision was made to allow a German army again, the next question concerned the form this new army should take.

The concept of *Innere Fuhrung*, which laid the basis for reform, was developed by Count von Baudissin and his colleagues between 1951 and 1956. The basic concept aimed toward an intellectual, political, and moral reform of the military. The only task still to be conferred on the soldier was the prevention of war through deterrence. Should the Bundeswehr be forced to fight, the battle would depend on the "thinking obedience," the initiative, and the conscious responsibility of each and every soldier. Such virtues could be achieved only if the soldier was considered a citizen in uniform and was granted—even within the military organization itself—as much as possible of the freedom that the soldier was committed to defend.

The creators of *Innere Fuhrung* acknowledged at most only slight differences between the military and industrial society. Increasing mechanization and the pressures toward specialization and intensified division of labor favored the process of assimilation and increasingly leveled the differences between the two. The encroachment of technology into the military organization was considered the cause of a continuing trend in which the military was becoming more civilian-like, and in which the differences between military and civilian work often disappeared completely. Without any great difficulty, the Bundeswehr could be compared with a public utility; its product was security. Therefore, the soldier's job could no longer be considered a job *sui generis*, but only as an occupation like any other.[4]

Much of the *Innere Fuhrung* concept was realized between 1956 and 1958 within the context of military legislation. The military constitution was later incorporated into the Basic Law of 1949.

The legislators accommodated the concept of the citizen in uniform insofar as the civil liberties of the individual were guaranteed to remain in effect, even within the military service. The soldier could be restricted in exercising civil rights only in a few specifically designated cases. No qualitative differences existed between the citizen and the soldier. The soldier might be politically engaged, vote, and run for election like any other citizen. The soldier had the right to join professional organizations and unions and to work for their goals. Disciplinary offenses were to be punished according to military disciplinary regulations, but the decision of the military disciplinary authorities would be subject to complete review by civilian judges. Independent military justice and the former military courts would not exist. Should a soldier become liable for prosecution (as through unauthorized absence, desertion, insubordination, mutiny, or mishandling of subordinates) the case would fall under the jurisdiction of the general criminal law.[5] Classified discharge in any form, such as honorable or dishonorable, would not exist.

Other aspects of the *Innere Fuhrung* were more difficult to implement, especially those parts that could not be instituted simply by legislation or regulation. Foremost among these were the reform concepts for daily leadership practices in the armed forces, which were impossible to transfer into concrete instructions and systematic learning aids that would be applicable in practice. Therefore, they remained primarily political slogans and general goals for a long time.

Above all, Baudissin and his colleagues were confronted with fierce criticism of their attempts to equate military and civilian work. Some objections were based on the incompatibility of military and civilian existence. In the 1960s, an entire series of literature appeared, condemning "Baudissin and his mistaken speculations." Many authors demanded a retreat from the "Baudissinist concept," which had proved ineffective because it had constructed an "unsoldier-like army."[6]

In the 1970 White Paper, the federal government attempted to describe the "new type of soldier." Definite functional characteristics were attributed to the soldier's duty; these, however, could also be found in the descriptions of other professions and occupational groups. In a clear trend toward occupationalism, soldiers were regarded as employees just like policemen, firemen, workers on an assembly line, and technicians.[7]

In January 1970, before the publication of the White Paper, nine theses on the occupational concept of an officer appeared under the title "The Lieutenant 1970." These theses had been composed by a group of lieutenants from the Hamburg Army Officers School. Thesis Eight read: "I want to be an officer in the Bundeswehr who demands a clear separation between duty time and leisure time because I view my occupation as a responsible and demanding 'job.' "[8] That an officer described his service as a job was unprecedented in the German military; many traditionalists considered this description close to sacrilege.

The nine theses of the Hamburg lieutenants were the prelude to a series of public exchanges from the ranks of the military. Opposing theses were formulated and published. The participants in the discussion were the military interest groups: the German Armed Forces Association (DBwV:

Deutscher Bundeswehr-Verband eV.) and the Union for Public Utilities and Transportation (OTV: Gewerkschaft Offentlicher Dienst, Transport und Verkehr).

Both organizations had competed to represent military personnel's interests, the DBwV since 1956 and the OTV since 1964. The DBwV was more internally oriented; it advocated the concept of the professional soldier. The OTV, as a union for all employees in public service, was more externally oriented; it represented a more civilian-like concept of the military vocation. The soldiers' sympathies can be determined clearly from the membership figures of these two organizations: the DBwV has nearly 300,000 members, representing more than 60 percent of all soldiers in the Bundeswehr, including conscripts.[9] The OTV had 1,500 soldiers as members in 1975; a more recent figure has not been published. The membership is actually very small, representing possibly less than one percent of all military personnel.

THE CURRENT DISCUSSION ON THE MILITARY VOCATION: THE "THIRD" APPROACH

The discussion about a vocational model and an occupational concept for the military continues up to the present. In January 1985, the Minister of Defense ordered once again that an officer's model be worked out, in which—as in earlier undertakings—the "common aspects among all officers" should be presented. The results of this study should produce an orientation aid for officers and officers in training, serve as a basis for training and education, and contribute to internal integration. Concurrently, this study should clarify to the public those characteristics of the officer's profession that are inherent to his duties.

The current discussion, however, has lost much of the fervor of the debates in the 1960s and 1970s. In addition to the traditional and the reformist orientations, a third approach has emerged—the pragmatic.[10] (See Table 11-1.) The national discussion on the military vocation can be clarified schematically, as shown below, by applying these three approaches.

On the ideological level and according to the official version, the debate continues between traditionalists and reformists concerning a vocational model for the military. In daily practice, however, the pragmatic approach has come to dominate; this approach is a synthetic result of the ambivalence present in the discussion in the Federal Republic, which could be found even at the time of German rearmament.

Although the DBwV still officially advocates the more professional vocational model, its demands for improvement of the soldier's economic and social position are often raised and justified by referring to developments in civilian employment. The association is pursuing a form of politics à la carte; for example, it has demanded legislation stipulating duty hours for soldiers according to the 40-hour week of the Public Service. In this dispute, which

TABLE 11-1. Three Conceptions of Military Service in the Federal Republic of Germany

Aspect	Traditionalist	Reformist	Pragmatist
1. Purpose of the Bundeswehr	Defense of the country against its enemies (use of force)	Deterrence (and defense) within the NATO alliance (prevention of armed conflict)	1. Deterrence 2. Defense 3. Aiding the civil population at home and abroad
2. Soldier's task	To fight and win (war service)	To prevent war and keep the peace (peace service)	To deter and, in a defensive emergency, to fight
3. Nature of the military	The military is an order-like community of fighting men/warriors	The military is part of the public service (insurance against injury from the outside)	The military is a large, bureaucratic-technical organization with functionally determined characteristics
4. Soldier's ethos	The soldier is a warrior	The soldier serves the cause of peace	The soldier must be prepared to fight in order to not be forced to fight
5. Vocation concept	A profession *sui generis*, unlike all others	The work is a job like any other	A profession with comparable and incomparable characteristics
6. Vocational motivation	Fatherland, honor, duty, Western culture, anti-Communism, Eastern enemy (community)	Freedom, law, human dignity, democracy (society)	Own interests, life-style, upward mobility, career, salary (individual)
7. Relationship to civilian society	Soldiers have a special status (incompatibility)	Soldiers are citizens in uniform (convergence)	Soldiers are citizens among citizens, but with special duties and restricted rights
8. Uniform	Uniforms are honorable attire	Uniforms are work clothes	Uniform are battle outfits and work clothes
9. Soldiers and politics	Soldiers should remain nonpartisan and feel committed to community good (servant of the state)	Soldiers should exercise civil rights and be involved in politics and society (citizen of the state)	Soldiers should exercise civil rights but remain nonpartisan
10. Interest groups	DBwV ("professional")	OTV ("union life")	Favors DBwV but also accepts OTV
11. Salary	According to rank	According to performance	According to rank with bonuses for performance
12. Duty hours	A soldier is always on duty	40-hour week similar to the civil service	Overtime pay or compensation in time off for over 40 hours a week

started in 1974–75, the customary positions have been reversed. The government, which usually favors a more civilian-like vocational model of the soldier, refers to the special features of military service that result from its task of defending the country: "The daily duty hours of servicemen are governed by military requirements. . . . A rigid scheme, such as the 40-hour working week of the Public Service is impracticable in the armed forces."[11] On the other hand, the DBwV, which usually advocates the professional vocational model, demands in principle that soldiers receive the same treatment as other employees in the civil service:

> What works for hundreds of thousands of civil servants, white-collar workers and workers employed with the Bundeswehr, as well as for the border guard troops, the police, the customs service, the postal service, the railroads, the health service, the emergency services, and the public utilities, cannot be withheld from our armed forces under peaceful conditions since they are also an important element of our welfare state.[12]

The military leadership of the armed forces continues its attempt to maintain the "unity of the military profession" under all circumstances. Therefore, the "ideology of the universal applicability" ("an officer can do anything!") is followed internally.

Officially, the adherence to the concept of a "unity of the military profession" leads to an overemphasis of the soldierly elements of the vocation. The recruiting agents of each branch of the military, however, cannot deny that many of the military skills no longer have much in common with soldier's work as it is traditionally conceived. In any case, they promote the Bundeswehr as a secure job for a limited period of time with primarily civilian-like occupational skills.

INSTITUTION VERSUS OCCUPATION: SOME EMPIRICAL FINDINGS

Legitimacy, Societal Regard, Self-Image

Military service in the Federal Republic of Germany has never been undisputed, and military life has never been easy. The rearmament of the country, which the public contested fiercely in the beginning, was soon generally accepted as inevitable, and the Bundeswehr was accepted as a necessary evil. The majority of the citizens also quickly reconciled themselves to compulsory military service. The population has grown accustomed to the uniform, but whoever enlists voluntarily in the Bundeswehr experiences repeated questions about this occupational decision. Unlike members of other occupations, soldiers must be prepared to explain the purpose of their service and to justify their existence.

All this has nothing to do with the actual fact that one is a soldier. The soldier is integrated into society and is accepted for the most part, despite relatively little social prestige.[13] In a similar sense, the Bundeswehr is also accepted as a governmental institution; it has the citizens' trust. The population, however, has trouble identifying with the task of the armed services, and as a result every soldier feels pressure to justify the decision in favor of a military career.

Central Europe is the most militarized region in the world. No other area contains so many soldiers and so much weaponry, including nuclear and chemical arms, in such a small territory.

The densely populated Federal Republic has a long common border with the Warsaw Pact countries—nearly 1,100 miles. Eighty percent of the Soviet land forces are stationed on this border.[14] In the case of an armed confrontation between East and West, the Federal Republic would become the battlefield, with incalculable consequences even if "only" conventional weapons were used. Soldiers then must calculate that their armed engagement would ultimately destroy precisely what they are supposed to protect.

To confront the doubts concerning the justification of soldiers, the political leadership and the military command of the Bundeswehr have developed a "legitimacy strategy." This strategy includes, above all, a new military ethic based on the fundamental concepts of *Innere Fuhrung*; the legal stipulation of a soldier's duties as deterrence and defense; the strong emphasis on the idea that the military service is a "service for peace"; and finally, the reference to membership in the NATO alliance as part of a "common lot with the free peoples," to reassure the citizens of the Federal Republic that they would not be alone in the case of armed conflict.

The new military-sociological literature of the Federal Republic includes much discussion about a legitimacy crisis in national security policy (and therefore also in the armed forces).[15] As a result of the NATO double-track decision of December 1979 and the stationing of Pershing II and cruise missiles in West Germany, which began in 1983–84, doubt about the logic of current strategy has increased. Visible evidence of this doubt is the emergence of a peace movement and its resonance in parties, churches, unions, universities, and especially the mass media. The Bundeswehr has not remained unaffected by this explosive public debate, especially because it involves the raison d'être of armed forces in the nuclear age. In this vein reports are "that it is the younger officers who are beginning to doubt the logic of a defense in a nuclear war."[16] This uncertainty is unlikely to be so widespread that one can speak justifiably of a "crisis atmosphere among the officer corps," especially because the public debate has since subsided.[17] However, the frequently intense societal criticism regarding the established policy of national security and the role of the armed forces has certainly been detrimental to the feelings of self-worth among many career and enlisted soldiers.

Soldiers need to feel confident that they have accepted an important and useful task for others and for their country. From such confidence, they obtain their occupational identity and their feelings of self-worth. The results of a

1978 survey of Bundeswehr officers revealed "a self-image of a conservative occupational group."[18] The responding officers acknowledged little significant difference between their work and comparable civilian professions (teacher, manager, engineer, technician), but they ascribed to themselves considerably different attributes and occupational motives. In comparison to other occupational groups, the officers described themselves as "unchanging," "rather skeptical of reform," "loving the fatherland," "patriotic," and "discipline-conscious." They considered themselves and their comrades to be less "career-oriented" and not "interested primarily in income" as members of other professions.[19] The survey also revealed that the officers' self-image and their image among the general population agreed for the most part.

The results from 1978, however, cannot be easily applied to the present situation. At that time there was no public discussion on national security policy or on the German armed forces. Strategic issues were strictly the business of a small circle of experts. Reliable insights into the present situation could be obtained only through recent empirical studies. Even so, several reference points guardedly indicate the current situation and further developments.

Thomas Ellwein pointed out in 1981 that the military establishment tended to react to societal criticism "by retreating into a fortress mentality created by shaping a special image of soldiers (and) by cultivating certain behavior which emphasizes distance from society."[20]

An unpublished 1982 study from the German Military Reserve Institute concerning the vocational self-concept of staff officers (who constitute nearly two-thirds of the career officer corps) revealed that a feeling of uncertainty had resulted from the societal criticism that was first voiced in 1980; the willingness of officers to justify their existence continually and publicly had also decreased. Complaints were made about the lack of an intact professional-vocational concept, the increasing importance of material interests and motivation (such as salary and leisure time) in daily military life, the increasing tendency toward a job mentality, and the expanding bureaucracy.

In summary, the specific military-political situation of the Federal Republic has hindered the career soldier and the longer-serving volunteer from creating a generally accepted and uncontested vocational model and occupational concept of their profession. Among the younger officers, who have experienced a different sort of socialization from their older comrades because of the obligatory civilian-oriented education introduced in 1973, this situation has led to problems of identification and motivation. (Reporting details about the scope and degree of such a development is not possible because necessary studies have not been conducted.) Older officers regard themselves as on the defensive, which has led them to maintain their professional self-confidence and self-worth by reverting to conservative values and traditional military virtues. This conservative trend can be discerned in the recent statements of

TABLE 11-2. Composition of the Bundeswehr, 1984

Personnel	Total numbers of service personnel	Officers	Non-commissioned Officers	Privates Airmen, Ratings
Soldiers on active duty	471,448	42,039	141,800	287,609
(in %)	(100)	(100)	(100)	(100)
Regulars	65,514	31,190	34,322	2
(in %)	(14)	(74)	(24)	(–)
Short-term and long-term volunteers	184,346	9,650	107,474	62,222
(in %)	(39)	(23)	(76)	(23)
Conscripts	221,588	1,199	4	220,385
(in %)	(47)	(3)	(–)	(77)
Soldiers trained under the vocational advancement plan	9,244	134	8,984	126
TOTAL PERSONNEL	480,692	42,173	150,784	282,735

Source: MOD, Bonn

the political and military leadership and in the increasing number of relevant publications in the military press.[21]

Composition, Role Commitments, Socialization

As of 1984, the military personnel of the German armed forces totaled 471,448, of whom 65,514 (13.9 percent) were career soldiers. 184,336 (39.1 percent) were short-term and long-term enlisted soldiers; and 221,588 (47.0 percent) were conscripts (see Table 11-2). Divided by rank, 8.9 percent of the military personnel were commissioned officers, 30.1 percent were noncommissioned officers, and 61 percent were rank-and-file soldiers.[22]

Therefore, the Bundeswehr is composed of three status groups who are in uniform for very different reasons. The career soldiers have chosen the military as a life vocation. The enlisted soldiers have committed themselves to the military temporarily (two to 15 years); for them, the Bundeswehr is merely part of a professional career. The conscripted men became soldiers because they were legally required to do so. This scheme of varying motivations determines the identification with the tasks, values, and norms of the military.

Career Soldiers

Recruiting enough career soldiers, either commissioned or non-commissioned officers, was never difficult. An NCO may become a regular soldier only when he has reached the rank of sergeant and the age of 25. A commissioned officer may become a regular officer only upon the successful completion of his education and upon promotion to first lieutenant. Thus the achievement of career-soldier status is preceded by a process of selection (both by the individual and by superiors) stretching over a period of years. This process guarantees that the only applicants for the corps of regulars who will be accepted are those who feel a strong personal association with the values and norms of the military.

Very little information exists about the socialization of career officers and enlisted personnel, but the long-term enlisted people and the career officers identify to a greater degree with things military and prove to have a "conservative basic attitude to the point of being traditional."[23] The previously mentioned study on staff officers, which has yet to be published, reveals a pronounced "military self-image." These career officers felt securely attached to and integrated into the hierarchical structure of military organization, which they fully accepted. They were skeptical of increasing specialization, which would force the soldierly aspects into the background. They believed that technology endangered the classical hierarchical structure of the military, especially in the air force, where the contrast between managers/technicians and soldiers is supposed to have created communication problems.

Short-Term and Long-Term Volunteers

In contrast to the case of career soldiers, the Bundeswehr has never been able to achieve the authorized personnel strength stipulated for short-term and long-term enlisted personnel. Even today it is lacking approximately 15,000 and the vacancies must be filled with conscripts.

Enlisted soldiers can commit themselves to a term of service ranging from a minimum of two years to a maximum of 15 years. Currently the average term of service among commissioned officers is 12.1 years, among NCOs 7.7 years (4 years, 38 percent; 8 years, 23 percent), and among the rank and file 4.2 years (2 years, 28 percent; 4 years, 66 percent).

In the case of the rank and file and the NCOs, applicants are assigned a trial period of six months. Only after that time is a term of service stipulated, which at first does not usually exceed four years. Even those who wish to enlist for a longer period of time—perhaps for special training—receive a contract for four years at first. The applicant is awarded the desired term of service only after the training is successfully completed. The term of service is therefore agreed upon by contract only if the soldiers have proved during training that they are capable of achieving NCO rank and later that of sergeant. The same holds true for officers in training. Their term of service is stipulated only upon the successful completion of their officer's examinations. Therefore, no one needs to risk serving a full term without having successfully completed the necessary training. At the same time, these procedures account for the fact that attrition (separation of personnel before completing the term of service) is not a problem.

A series of studies has been conducted on the social origin and enlistment motivation of short-term and long-term volunteers. The studies have shown that approximately half of such young men come from rural communities and small towns and that the majority of them are skilled workers, artisans, or technicians. They are interested in military service primarily because of the financial aspect and for motives connected to their civilian profession. The unfavorable state of the employment market also plays a role.[24]

Volunteers who apply for officers' careers are primarily interested in the educational program introduced in 1973. This program leads to an academic diploma and is obligatory for applicants who enlist in the service for a minimum of 12 years.

In a representative sample of young men considering enlistment in the armed forces in 1983, 78 percent viewed such a commitment to the Bundeswehr as a way to learn much that would be useful to them later in a civilian profession, and 22 percent said that they were interested above all in "being a soldier."[25]

The Bundeswehr must submit to their expectations and wishes. Military training is already oriented toward comparable civilian occupations, and the soldier can even take a civilian final examination during the term of service. Short-term and long-term enlisted personnel have the right to vocational training either during or after the term of service; the duration of such training depends on the length of time served. The Bundeswehr maintains 28 professional schools for this purpose, which has made it one of the largest and most efficient institutions of adult education in the Federal Republic.

The effect of military socialization on short-term enlisted soldiers is rather slight. They remain much more oriented toward civilian life than do the "lifers," the career soldiers. They view their temporary military occupation as one profession among many, and the differences between their own self-image and their public image are small. Dillkofer and Klein have come to the conclusion, based on the noncommissioned officers of the Bundeswehr, that the differing orientations of career and short-term soldiers lead to such specific differences of attitude and opinion "that any comprehensive vocational model of the 'NCO' must be placed in doubt."[26]

Conscripts

Compulsory military service is not popular, especially not among those affected by it directly. In the 1980s, on the average, approximately 50,000 young men a year have applied for recognition as conscientious objectors. Compulsory military service is not considered an intolerable demand; however, among young men there "prevails a reasonable, pragmatic attitude towards compulsory military service which is hardly feigned or ideologically tainted."[27]

The application for recognition as a conscientious objector is not always a renunciation of compulsory military service, the Bundeswehr, or defense readiness. In making such a decision either for or against compulsory military service, the personal cost-benefit calculation and the sober consideration of the advantages and disadvantages of each choice also play important roles. The effects of compulsory military service on the attitudes and behavior of conscripted soldiers are hardly discernible. Conscripts remain civilians in uniform; they remain skeptical or negative about the military educational proceedings. The Bundeswehr does not attempt to restrict, let alone break, the contacts and connections between conscripts and civilian society. As a matter of principle, conscripts are called up to garrisons near their homes. The majority of the conscripts spend their weekends at home; a large number of the conscripts stationed close to home, whose barracks are located virtually in their front yard, have "home-sleeper" permits: they appear every morning at their workplace in the barracks and go home every evening, just like civilian employees.

CONCLUSION

In contrast to the armed forces of other countries, whose military history developed without interruptions, the Bundeswehr is a new creation. It was the declared will of the lawmakers to integrate the new German armed forces as much as possible into the industrially developed and democratically constituted society of the Federal Republic, to prevent military life from developing again independently of its society, and to prevent the new military from becoming once again a "state within a state." The citizen and the soldier should be, in the words of Graf von Baudissin, "only two states of aggregation of one and the same citizenship." Conscription was conceived as a clamp holding together the armed forces and society.

The fathers of the Bundeswehr were clearly influenced by the idea of an occupational military. The military was conceived as an "institution of public services"; the military profession was viewed as a profession like any other; the working conditions and duty requirements of the soldiers were to be made as similar as possible to the conditions of civilian employment and especially similar to those of the civil service. Different regulations should exist only in cases where the functional characteristics of the military required them.

The reformers' ideas were largely realized in the military legislation. In many respects, the Bundeswehr became the prototype of the occupational military. Nevertheless, historical experience and ingrained personal traits cannot simply be supplanted by other concepts; the inertia of traditional concepts concerning soldiers and the military life have always been strong, and they are so now.

Even during the period of establishing the Bundeswehr, the reformers were forced to reduce their expectations. Count Baudissin, for example, proposed equipping soldiers only with a combat outfit and fatigues; these would be the soldier's work clothes. The actual uniform could be dispensed with easily because, after all, no other vocational group has such a parade uniform.

If one uses the early years of the Bundeswehr as a reference, the armed forces once again have moved ideologically closer to the institutional military. This trend becomes obvious in the revival and intensification of military tradition and soldierly folklore and in the reintroduction of flags and medals. Today one can find the tendency, especially among career soldiers, to revert to the traditional self-image, to genuine "soldierly" virtues, and to the ideal of "selfless service for the community."

Exactly the opposite tendency can be observed among the conscripted soldiers. Their civilian orientation, which has always been strong, is increasing, and the rejection of all things specifically military is growing. Convincing conscripts of the necessity and reasonableness of the exercise and maintenance of certain military forms (formal exercises, roll call, obligation to salute) is increasingly difficult.

Hence, two opposing trends coexist within the Bundeswehr. The cleavage runs horizontally through the ranks of the noncommissioned officers and separates the higher ranks (A7 to A10) from the lower (A5 and A6). Between these two is an obvious border: "Here collide two different worlds of ideas."[28] Whereas the older sergeants participate in the military establishment's traditional tendency to separate itself from the civilian world, the younger NCOs and the conscripts tend to favor civilian forms and elements. Thus, the Bundeswehr could be called a "divided army," as Charles Cotton calls the Canadian armed forces.[29] Nevertheless, I emphasize again that until now this evaluation of the Bundeswehr has been possible only on the basis of secondary research.

NOTES

1. For a summary review, see Dietmar Schossler, *Militaersoziologie* (Konigstein: Taunus, 1980), pp. 159–193.
2. Charles C. Moskos, "Institution versus Occupation: Gegensatzliche Modelle militaerischer Sozialisation," in Ralf Zoll, ed., *Sicherheit und Militaer* (Munich: Westdeutscher Verlag, 1982), p. 199–211.
3. See, for example, Detlef-Lothar Baehren, "Beruf and Arbeit des Unteroffiziers in modernen Armeen," in Paul Klein, ed., *Das Strapazierte Rueckgrat, Unteroffiziere der Bundeswehr* (Baden-Baden, 1983), pp. 53–88.
4. On the preparation of *Innere Fuehrung* and the conceptual content, see, for example, Dietrich Genschel, *Wehrreform und Reaktion. Die Vorbereitung der Inneren Fuehrung 1951-1956* (Hamburg, 1972); Wolf Graf von Baudissin, *Soldat fuer den Frieden Entwuerfe fuer eine zeitgemaesse Bundeswehr*, edited and introduced by Peter von Schubert (Munich, 1969).
5. The establishment of special military criminal courts is stipulated by article 96, para. 1, of the Basic Law only in states of defense.
6. Included among these publications are Heinz Karst (1964), Winfried Martini (1960, 1964),

Friedrich August Freherr von der Heydte (1965), Hans-Georg von Studnitz (1967), Adolf Reinicke (1968), and Friedrich Dopner (1969).

7. *White Paper, 1970, The Security of the Federal Republic of Germany and the Situation of the Federal Armed Forces* (Bonn, 1970), pp. 115–116.

8. Carl-Gero von Ilsemann, *Die Bundeswehr in der Demokratie. Zeit der Inneren Fuehrung* (Hamburg, 1971), p. 328.

9. According to the DBwV's own statistics, approximately 80 percent of the regulars and volunteers have joined the association. This rate of "organized" membership has not been reached even by the Public Service, a group that tends traditionally to be organized.

10. Also compare the results obtained by Werner Kriesel from the SOWI research project, "On the discussion concerning a vocational model for the officers of the Bundeswehr." Unpublished manuscript (Munich, 1986).

11. *White Paper 1979, The Security of the Federal Republic of Germany and the Development of the Federal Armed Forces* (Bonn, 1979), p. 255.

12. *"Der Mensch im Mittelpunkt" — Der Deutsche Bunderswehr-Verband zur sozialen Lage der Soldaten* (Bonn, 1985), p. 16.

13. For a single comprehensive survey, see Ralf Zoll, "Militaer und Gesellschaft in der Bundesrepublik: Zum Problem der Legitimitaet von Streitkraften," in R. Zoll, ed., *Wie integriert ist die Bundeswehr?* (Munich: Westdeutcher Verlag, 1979), pp. 41–76.

14. Wolfgang Altenburg, *Die Bundeswehr: hier und heute* (Koln, 1985), p. 7. General Altenburg is the chairman of the NATO Military Committee.

15. See especially Wolfgang R. Vogt, ed., *Sicherheitspolitik und Streitkrafte in der Legitimationskrise* (Baden-Baden, 1983).

16. Paul Klein, "Berufs-und Zeitsoldaten in der Identifikations und Motivationskrise," in Wolfgang R. Vogt, ed., *Sicherheitspolitik*, p. 301.

17. Ibid., p. 304.

18. Michael Bohrer, *Offizier der Bundeswehr: Selbst—und Fremdbild, Berichte des Sozialwissenschaftlichen Instituts der Bundeswehr,* Heft 32 (Munich, 1983), p. 5.

19. Ibid., p. 49 f.

20. Bundesministerium der Verteidigung, ed., *Soldat und Gesellschaft* (Koln, 1981), p. 30.

21. See, for example, Manfred Wörner, "Zum Bild des Offiziers," in *Soldat und Technik* 8 (1983): 407–411. M. Wörner is the present minister of defense.

22. In addition, approximately 180,000 civilian employees support the armed forces by performing tasks that are not strictly military.

23. Heidelore Dillkofer and Paul Klein, *Der Unteroffizier der Bundeswehr II. Rekrutierung, Berufszufriedenheit, Selbst-und Fremdbild, Berichte des Sozialwissenschaftlichen Instituts der Bunderswehr,* Heft 21 (Munich, 1981), p. 136.

24. Instead of many others, see: Dillkofer and Klein, *Der Unteroffizier..*

25. Bundesministerium der Verteidigung, Informations-und Pressestab 3, *Junge Maenner und Bundeswehr: Einstellungen und Dispositionen* (Bonn, 1983).

26. Dillkofer and Klein, *Der Unteroffizier*, p. 203.

27. Ekkehard Lippert, "Sozialization," in Ralf Zoll, Ekkehard Lippert, and Tjarck Roessler, eds., *Bundeswehr und Gesellschaft. Ein Woerterbuch* (Opladen, 1977), p. 294.

28. Dillkofer and Klein, *Der Unteroffizier*, p. 203.

29. Charles A. Cotton, *The Divided Army: Role Orientations Among Canada's Peacetime Soldiers* (Unpublished Ph.D. dissertation, Carleton University, Ottawa, 1980).

XII

France

BERNARD BOËNE

In defense matters, France occupies an unusual position in many ways. Although it is a member of NATO, it does not participate in the organization's military side (from which it withdrew in the mid Sixties). It possesses a small, independent nuclear deterrent (now entering its third decade) based on a strategic concept of massive countercity retaliation. Yet, it has retained a conventional armed force that can be regarded as comparatively large given the size of its territory (1/16 of the continental United States), its population (55 million), and substantial levels of friendly forces to the east and north, protecting it from direct contact with the threat. This force is roughly similar to that of West Germany, a nonnuclear, frontline country.

Since the final retreat from empire in 1962 (following nearly two and a half decades of war successively in Europe, Indochina, and North Africa), France has been at peace with the world. The recent operational experience of French troops has been slight in absolute terms, though considerable by contemporary European standards; most of the French military activity in the last 20 years or so has taken the form of small-scale operations in Africa and the Middle East in aid of friendly governments, in defense of French or Western interests, or on various peacekeeping missions. These operations include "surgical" interventions (Gabon in 1964, Chad in 1968 and 1987, Zaire in 1978), cease-fire supervision, interposition, and military assistance. Although the French have suffered tragic moments (such as the Beirut suicide bombing in late October 1983 in which more than 50 French paratroopers died, including conscripts), they never aroused any considerable public interest. They have been marginal at best, though obviously important to the handful of marine and foreign legion battalions earmarked for that kind of mission overseas and of great symbolic value in the army far beyond the actual numbers involved.

With the advent of nuclear weapons and the withdrawal from NATO's military command structure, French strategic interests have centered mainly on the security of the national "sanctuary" through nuclear deterrence, with protection of lifelines and the external security of the few remaining overseas dependencies as secondary preoccupations. This position was not without its ambiguities in the late Sixties and throughout the Seventies: go-it-alone military policies were not easily reconciled with political obligations to partners under the Atlantic Alliance. The use to which French troops stationed in Germany would be put in time of crisis or war, and the fielding of the first generation of tactical nuclear weapons (Pluton), raised persistent questions about the desirable level and scope of military cooperation with political partners. This was the crux of the 1970s: controversy over advance planning for a possible French contribution to NATO's "forward battle" in Central Europe. Although it was advocated very cautiously by some high-ranking military officers (including the then defense chief of staff) under Valéry Giscard d'Estaing's conservative government, and although it apparently caught the imagination of the top political leaders (including the president himself), for a variety of reasons such a contribution never became official policy. In the early Eighties, however, under President François Mitterrand's socialist rule, France—virtually unaffected by peace movements of the type prevalent in other Western European nations—came out strongly in support of NATO deployment of theater nuclear weapons. It is now in the process of fielding a 47,000-strong *Force d'Action Rapide*, designed for the intervention on short notice in various parts of the world, including Central Europe. This rapid-deployment force, which comprises an airmobile and air assault division, could not operate entirely in Central Europe without advance planning and cooperation with U.S., British, and German forces.

Despite appearances to the contrary, political, military, and cultural logic lies behind this situation. Twenty-five years ago, France—then under General Charles de Gaulle's strong leadership—resolved to put an end to a quarter-century of humiliation, frustration, and weakness. The loss, or rather the eventual renunciation, of its imperial position made France all the more determined to continue playing a world role worthy of its past, that is, a role not strictly proportional to its present demographic and economic weight. Having failed to convince the United States that such a claim was based on reasonable grounds, France felt that it had no alternative but to build nuclear weapons of its own, leave the Alliance's military arm, and pursue independent policies whenever the relations between the two superpowers allowed enough maneuvering room; when the chips are down, France usually proves a most faithful ally of the United States. The Gaullist tradition, which seems to enjoy a strong popular consensus, has continued until today.

The military logic resides in the fact that possession of nuclear weapons today normally turns a country into a "political-strategic island," to borrow General Lucien Poirier's phrase.[1] This new situation entailed modernization of the traditionally smaller "services in blue" (the air force and the navy)—more accurately, of their strategic components and a few others besides. Because of budgetary constraints in the Sixties and early Seventies, the modernization of the army did not rank as a top priority, and the khaki service was left behind. This discrepancy produced what Michael L. Martin has described aptly as a sharply dualistic structure: for lack of sufficient resources, France based its defense concept at first on an all-or-nothing doctrine that provided functional roles for much of the navy and the air force, but deprived the large, archaic army of a meaningful part except, in its own self-conception, as NATO's second echelon.

Economic and military rationality cannot explain why no one thought of significantly reducing the army's strength. This is where the cultural logic takes over: no one dared to do away with conscription as we know it—a universal, egalitarian draft, continuously in force since 1905, which has been one of the traditional republican forms of civilian control.[2] Associated in the collective memory with revolutionary fervor and the glories of the Napoleonic wars, the army was used effectively until the 1950s as a vehicle of social policy to produce a sense of unity in what was then a geographically and socially heterogeneous country and to turn every young Frenchman into a citizen. In addition, a country that had been invaded partly or completely three times in less than a century and that was only beginning to come to terms with the subtleties of deterrence was understandably bound to feel more secure with large numbers of its own ground troops, even if poorly equipped, on its northeastern borders.

Nearly two decades later, things have changed substantially. The army, starting with the Fourth Military Programming Law (1976–82), has begun its technological modernization, whose attendant social-organizational effects are expected to diffuse widely in the not-too-distant future. Although conscription still enjoys a sizeable popular and political consensus, it is now undergoing a slow process of "detabooization," as Martin predicted correctly in his authoritative book *Warriors to Managers*.[3] Force levels are already being reduced. Thus, the French conventional armed force of the near future, leaner and probably meaner, is moving toward the goal of giving the president the option of trying to influence battlefield events in Europe, should the Soviets cross the Iron Curtain, before (or instead of) pressing the nuclear button as the attack reaches the national sanctuary. Even though this development does not herald a new gradualist approach to deterrence, much less a return to NATO's integrated military structure, it does reinforce the trend toward closer military cooperation with allies.[4] The important fact, from a sociologist's point of view, is that France's conventional forces, especially the ground forces, are feeling functional and relevant once again, and—however slowly—are regaining a sense of purpose.

THE FRENCH ARMED FORCES TODAY

In 1986, overall French military strength was approximately 500,000 on active duty. About half of those in service were conscripts, most of whom served in the army.[5]

One major development in the last few years has been the emergence of the French version of the draft-motivated volunteer. As of 1983, conscripts have had the option, either before or after induction, of extending their obligated twelve-month tour by a further four, six, eight, ten, or twelve months. This option gives them the right to choose not only the service, but also the specialty or arm in which they wish to serve and even the location (whenever feasible) of their unit. They also draw double the pay of an ordinary conscript in the first year, and $2\frac{1}{2}$ times as high for the remainder of their extended tour. So far the move has proved a moderate success: a little over 10 percent have chosen this option.

This is not the only change in the French military manpower system.[6] Despite a moderate decrease (by 9,000) in the number of career volunteer military personnel and a very slight increase in civilian employee strength (by 2,000), the number of draftees has declined sharply (by some 34,000). In other words, France is moving from the mass armed force of 25 years ago to the much smaller active-duty force that military sociologists predicted would come about in most Western nations. Much of that "progress" has been made by the army, which long seemed to be lagging behind the other two services in that respect. A further 5 percent reduction in overall force levels has recently been announced and will be implemented in the next five years.

Likewise, because of increased technical specialization, especially in the army, the trend toward less homogeneity and less role interchangeability has continued. Social representativeness is slightly less today than in 1977. Although conscripts retain an overall majority, that majority has considerably narrowed; if present trends continue, career and volunteer personnel will soon outnumber them. Conscripts themselves are less socially representative than they used to be: only 60 percent of the males in an age cohort are drafted into uniform today. Five percent join the forces voluntarily, 3 percent opt for a civilian form of national service, 0.3 percent are conscientious objectors, 6.7 percent are exempted for family reasons, and 25 percent for medical reasons (a figure that compares with less than 10 percent in the 1950s). Moreover, 6 percent of those inducted into the military are discharged for medical reasons in the course of their obligatory 12-month tour. If current trends persist, the discrepancy between reality and the principle of universality and equality of military draft obligations will eventually undermine the legitimacy of conscription as we know it. However, if the draft obligation is limited to a six-month tour in accordance with one of President Mitterrand's 1981 campaign promises, more options and incentives will have to be provided to attract more draft-motivated volunteers, in order to compensate for the loss of ordinary drafted manpower. The likely result will be a further decline in social representativeness. In any case, we seem to be on the threshold of drastic changes in France's military manpower system, whether in the immediate or the medium-term future.

The decline of the mass armed force has also become more perceptible in regard to mobilization plans. In time of crisis or war, the reservists (all conscripted with the exception of officers over the age of 35, who are retained on a voluntary basis) would just double the present active-duty force, whereas they would have more than trebled the much larger active-duty force of 25 years ago.

Quality, especially in the army, and retention (mostly restricted to sophisticated navy and air force technical specialities, strongly influenced by a tight civilian labor market) were distinct problems in the early and mid-Seventies among volunteer recruits. The situation has improved dramatically since then: quality, which is now somewhat above the civilian average (in educational levels, if not in mental aptitude), has risen very sharply, owing probably to much higher levels of unemployment outside; to the rise in the relative social and economic status of career military and volunteer service members, which became effective in the late Seventies; and possibly to the general increase in the population's educational standards. For much the same reasons, the selective retention problems mentioned above are reported to be less acute today. (In fact, retention—due to cultural factors—is generally so high that the real problem is how to induce aging supernumerary senior officers and NCOs to leave.)

Barring drastic alterations in the draft law and the economic outlook, no changes are expected in recruitment, retention, and compositional characteristics in the near future. If, however, the draft should be curtailed and the economic environment should improve, as appears likely at least in the longer term, what the French military, notably the army, will look like in five or 10 years is anybody's guess. In particular, the widely postulated "natural ceiling" in the supply of traditional volunteer enlistments will have to be tested. (A general responsible for recruiting volunteers in 1913 was reported as saying that no one would ever be able to recruit more than 18,000 volunteers a year in France. The present level is slightly above 20,000, but that is the number of places as defined by the defense budget, not the maximum potential number of enlistees.) Much will probably depend on the elasticity of the newly created supply of draft-motivated volunteer enlistments. Finally, the present low level of civilian and military female recruitment will probably have to be raised.[7]

From now on, military personnel analysts will be watching very closely for the series of ground-breaking policy moves that the future appears to hold.

INSTITUTION-OCCUPATION INDICATORS

After this brief summary of recent manpower trends and development, it is time to examine the various indicators used by Charles C. Moskos in his stimulating Institution/Occupation (I/O) thesis, with an eye to placing the French military services on an I/O continuum (if, indeed, a single continuum best fits empirical findings).

Legitimacy

Even a superficial glance will convince any lay observer that differentiating between the army and the other two services on this point, as on several others, is necessary.

Army culture has long had a very strong normative, combat-oriented, communitarian, and paternalistic-authoritarian content, whose pervasive influence can be felt at all levels among career and volunteer service members. It is reinforced by the high rate of self-recruitment—cultural inbreeding—and by the easily perceptible desire among those who do not come from a military background for life-styles that differ from civilian life-styles. Draftees do not stay long enough to be fully socialized and, with a few exceptions, remain civilianized in outlook. In that sense, the facts do not support the supposition that conscripts would be more institutional and careerist and volunteers more occupational, at least in terms of norms and values. The influence of the army is all the more remarkable because, except for intense institutional socialization in military schools or during early basic training, there are few formal institutional controls; the system loosens as one gains seniority. This looseness, however, is at least partly compensated for by the fact that the promotion system tends to reward those who adhere most closely to core institutional values; those soldiers are also the ones who stay longest.

In light of this persistence of the traditional fighting ethic and of traditional collective self-conceptions, the 23 years of peace that have elapsed since the end of the Algerian war have created deep strains in the army. As General Saint-Macary once remarked, people in the land force (mainly officers) have worked hard to keep up appearances and to continue pretending to themselves that soldiering has remained basically unaffected by a strategy of nuclear deterrence and by the delegitimization of the use of armed force in Western nations.[8] Many have regarded the limited political-military fire brigade duties or peacekeeping missions that marine and foreign legion battalions have been called upon to carry out at irregular intervals in Africa or the Middle East as a poor substitute for the real thing. Among the other ground forces, the difficulty inherent in making peacetime training realistic, the drudgery of "standing, waiting and dry-running," and the time-consuming administration of the organization's everyday needs in the context of high conscript turnover have produced, in Charles Cotton's words elsewhere in this volume a "beleaguered warrior syndrome."

This situation was made worse in the 1970s by the feeling that the army was not getting its fair share of attention and budget resources, that its missions lacked strategic relevance, and that there was a growing separation between cherished military values and the civilian ethos of affluence. In brief, the army's institutional ethos, though still strong, has been undergoing a crisis of adjustment to a strange environment.

The future direction of current trends is unclear. On the one hand, events such as the tragic deaths of the French paratroopers in Beirut and of French soldiers in the United Nations forces in Lebanon in 1986 were hailed in traditional army quarters—in becomingly muted tones—as a timely reminder that notions of self-sacrifice to the point of giving one's life for one's country were not as hopelessly old-fashioned and devoid of meaning as some people were beginning to assume. These traditionalists have taken comfort from the revival of conservative values in Civvy Street in the last three or four years: patriotism, discipline, exertion, merit, short haircuts, and neckties are back, especially among the young, after an eclipse of nearly two decades. (An article that appeared in *Le Monde*, the liberal establishment newspaper, on March 7, 1985, commenting on a survey of young males shortly to be drafted into the army, went so far as to speak of a thirst for rigor, discipline, order, and espirit de corps.)

On the other hand, the comparative affluence of careerists and volunteers seems to contradict the old professed contempt for material wealth; more important, the much-desired (and much-feared) technological modernization—the prerequisite for new, strategically relevant missions—is well under way. It will be interesting to observe the organizational consequences, on the symbolic level, of increased specialization and the new patterns of relations induced by automated battlefield technology.

The organizational climate in the navy and the air force has long been much more relaxed than in the army. To begin with, air force personnel and sailors do not harbor metaphysical doubts about military survival in a peaceful environment, even though they have had no major fighting experience in four decades. (Their involvement in colonial wars was limited, and their participation in subsequent operations in Africa and the Middle East has been restricted mostly to logistical support, with only occasional air or naval air strikes.) The main reason for this lack of doubt is that peacetime exercises and general conditions of existence are sufficiently close to the reality of actual operations that they do not feel deprived of essential elements of their calling, and they develop positive attachment to their jobs, units, and service—and beyond. (In the navy, half the personnel serve on board ships at any given time; in the air force, even nonflying engineers and technicians bear heavy technical responsibilities.) They are also more civilianized in outlook than their army counterparts. For one thing, most of the skills for which both the navy and the air force provide training and experience have civilian equivalents. The traditional heroic fighting ethic is directly relevant only in small portions of these two services, and those concerned stay in such assignments for only brief periods, early in their careers. (Moreover, among naval and air force cadets, maritime calling and flying are much higher on the list of declared motivations than military calling or family tradition.) Some of their missions are civilian in character: for instance, 13 percent of the navy's activities are officially devoted to public service missions such as coast guard operations, surveillance of maritime transit, rescue operations, detection of oil slicks, and enforcement of European Economic Community (EEC) fishing regulations. For another thing, having undergone modernization much earlier, they tend to take the effects of specialization for granted and have long been accustomed to a functional rather than a normative type of integration.

In such a context, the social scientist need not be surprised to hear a navy officer, quoting an old naval saying, define himself as "a professional seaman first, a warrior only as chance dictates," or to find air force recruitment brochures with front-page titles reading: "The Air Force, A Major Public Sector Enterprise. . . ." However, a civilianized outlook is not the same as an occupational outlook, even if the navy and the air force are less institutional than the army.

In the early 1970s, the three services underwent a severe moral crisis. It was most acute in the army, as suggested above, but was also felt deeply in the "services in blue," where it translated into recruitment and retention shortages, especially among NCOs. The basic problems were low military pay and loss of prestige due to a perceived lack of functional and moral relevance in a world characterized by nuclear deterrence, economic affluence, and cultural (notably moral) permissiveness. The situation was deemed so serious that the government decided to intervene.

The remedy did not consist of the idea that the military person is basically a *homo economicus* like the rest of us and that labor market rates alone would take care of the problem; Milton Friedman has had disciples in France, but their influence has been minimal. The cure consisted of formally emphasizing that military service members are part of the national public service establishment and that, even if due allowance is made for a residual military uniqueness, the comparability of military and civil service should be made more apparent.[9] This idea was resisted by some army officers, who felt that it was an intolerable attack on their self-images, but it was widely accepted everywhere else in the military.

From then on, an officer could compare his lot with that of the other executive positions in the higher civil service (A grades: civil administrators in the central services of the ministries in Paris, college and university professors, professional civil service engineers, senior police officers), and an NCO could compare his position with that of people in the top brackets of the lower civil service (B grades: junior internal revenue officers, primary and junior high school teachers, technicians). The point was that since 1947 civil service pay rates had been higher than military rates at comparable educational and seniority levels, and—more important—given the central role played by the state in France for the past three centuries, the civil service enjoyed both a strong tradition and comparatively high social prestige.

According to legal theory of public servants, evaluation of public service performance for the most part is holistic and qualitative, unlike that of materially productive activities. Therefore, public service compensation is not granted by the state in exchange for particular skills or work capacities but, on the basis of the normatively appraised "dignity" of the duties or functions performed by the various categories of personnel, to afford them a standard of living in keeping with both the frugality of the public service and the respect it should inspire. Thus, the pay and prestige of military service members were raised in the mid- and late Seventies, while the military organization remained for the most part sheltered from the undiluted occupational influences.

Two things should be noted, however, before we leave this subject. First, military men and women are now among the best-paid public servants. Once the principle of parity between civil and military service was adopted, legislators decided that the unusual demands and constraints of military life (liability for 24-hour service, frequent moves, restricted civil liberties, lower age limits, risk to life and limb) deserved supplementary compensations. Second, pay rates in the lower civil service are generally higher than those for corresponding skill levels in the private economy, whereas compensation levels in the higher civil service are usually well below the salaries of private-sector executives.[10] In other words, NCOs are now better paid than most of their civilian counterparts, except for a few sophisticated technical specialists in high demand in industry, whereas officers' pay compares favorably with that of senior civil servants, but unfavorably with that of their private-sector peers. (Yet, if we consider the great proportion of officers who pursue long careers, the salary difference does not seem to trouble them much.)

Role Commitments

So far, this writer has not encountered complaints, official or otherwise, about a perceived lack of commitment of service members to their roles. In a qualitative sense, however, as could be expected, the army and the other two services differ substantially.

In the army, role commitments are broad and diffuse, in keeping with that service's strong normative orientations. This phenomenon is linked with a personnel management system based on arms and services: an infantry sergeant competes for promotion (and to a lesser extent, for postings) with other infantry sergeants, whether their military occupational specialties (MOSs) require difficult technical skills or only relatively unsophisticated combat skills. The same principle applies by and large to officers up to and including the rank of full colonel. This situation has traditionally ensured the primacy of the soldier's role over the specialist's role.

In the navy, most military personnel are managed on the basis of their technical specialties. Competition for promotion and assignments takes place among those within the same specialty, which induces the primacy of the specialist's role. (Characteristically, specialty insignia on petty officers' uniforms are often as conspicuous as grade insignia, if not more so.)

The air force stands halfway between army and navy: NCOs are managed on the basis of specialty (and the primacy of the specialist's role is such that some military duties, such as occasional guard duty, are reported to be unpopular), while officers are managed on the basis of broad functions (flying personnel, ground engineers, management, and security).

The army chief of staff recently ordered a feasibility study concerning a change from arms and services to specialties as the basis of the army personnel management system. This move reflects the increasing importance of high technology in the army, but at this time the probability that such a drastic move will actually take place is difficult to assess.

Positive attitudes toward identification with one's service and the 24-hour service liability are a matter of pride in all three services. In the last 10 to 15 years, those positive attitudes have become less visible, even if the norm remains strongly internalized. Symbolically, uniforms are now worn only rarely in off-duty hours outside military installations. (In the Paris area, even officers change into civilian clothes before leaving the workplace.) Recent questionnaire data has also confirmed a discernible trend toward a growing resentment of the traditionally wide scope of organizational control: interference of military life with personal and family life—mostly among career and volunteer personnel—is less well tolerated than it used to be. Moonlighting is strictly forbidden by law, however, and is practically unheard of.

Compensation

To return to the subject of pay, the basis of compensation in the military remains rank and seniority plus need, including family allowances, disability pensions, and regional cost-of-living adjustments, as in the civil service. The military services do take skill levels into consideration, but on their own terms: only approved and required skills, acquired mostly within the military framework, are remunerated, in the form of bonuses amounting to 10 or 20 percent of basic pay. Only a handful of officers at most take master's degrees, and later doctorates, in civilian universities.

In regard to types of skill, some activities involving risk to life and limb give rise to special bonuses; pilots and paratroopers, for example, receive a special "air service" allowance of some 30 percent of basic pay. On the other hand, the supply-and-demand status of a given type of skill on the labor market is not taken into consideration: the only known accommodations to the marketplace philosophy concern medical doctors and professional weapon-design engineers. In the specialty-based management systems of the navy and air force, however, those technical specialties in great demand on the outside generate higher pay differentials through the mechanical adjustment of promotion rates to a higher turnover.

With reference to the mode of compensation, the balance between noncash benefits and salary plus allowances or bonuses is overwhelmingly in favor of the latter. The incidence of subsidized housing or food is minimal, and the only perquisites are subsidized domestic railroad or air travel and subsidized uniforms (for NCOs only). Military retirement pensions (based on a 7 percent contributory scheme) are only very slightly higher than civil service pensions for a given number of years, and about 50 percent higher for a full pension than those of civilian private-sector workers or executives, who generally compensate by subscribing to voluntary complementary superannuation schemes.

The pay system seems extremely decompressed, mainly because nominal conscript pay is extremely low: 405 francs a month—10 percent of the private sector's minimum wage—for second-class privates. We must bear in mind, however, that they are housed, fed, and clothed.

In terms of basic pay, the equivalent of a sergeant-major around the age of 45 earns 23 times as much as a raw conscript recruit, nine times as much as a draft-motivated volunteer after the end of his 12-month obligatory tour, seven times as much as a traditional, longer-term volunteer recruit, about three times as much as a volunteer corporal, and twice as much as a sergeant. Among officers, the range of salaries is more limited: all things being equal, a brigadier general's pay is only three times as high as a second lieutenant's, and a lieutenant colonel's is only twice as high. The extent to which enlisted and officer pay scales overlap is comparatively limited; a sergeant-major in his middle forties earns about as much as a captain in his early thirties, and the pay of a young volunteer sergeant, after a year or two, is comparable to that of a second lieutenant.

Military pay levels are now considered satisfactory by all concerned, and civilians are beginning to look askance at military salaries. Even so, the new, comparatively affluent military service member is not quite the end of the story. We must also mention the superaffluent constabulary or peace soldier: a corporal serving today in Chad or in Lebanon earns as much as a major in continental France, and if he is a paratrooper, married with one child, as much as a full colonel in Paris. Sometimes the psychological consequences are profoundly destabilizing for low-grade volunteers who return later to low civilian wages.

Residence

Military installations in the continental United States are generally located away from cities, but French military installations, except for army training camps, have traditionally been situated in or near medium-sized towns or even large cities. When work and residence are separated, traveling to and from work can be a time-consuming process because of rush-hour traffic, but service members generally commute in the reverse direction, from the town centers where they live, to the outer suburbs, where military installations are located. Even the army, in many cases, has sold its old, historic barracks in garrison-town centers to build new, more spacious and functional ones on the outskirts. This situation, however, concerns only a minority, estimated at less than 25 percent—mainly married officers and NCOs. Draftees and low-grade volunteers live on base, if only for obvious economic reasons, except for a few who are married or assigned to units in their home towns. Bachelor NCOs and junior officers generally find it advantageous to live in the sergeants' or officers' mess, where they are looked after materially at prices that are generally much lower than in the civilian sector.

In other words, separation of work and residence does not pose much of a problem today, except possibly in the Paris area, where geographic circumstances often make life very uncomfortable for civilians and military alike.

Spouses

A 1983 study based on some 100 interviews with army officers' and NCOs' wives yielded the following findings.[11] Gainful outside employment is growing more common among the wives of junior NCOs and officers, not so much for economic reasons, but either because they wish to avoid the boredom of a traditional military housewife's life or because they have definite career plans and expectations, even though their husbands' nomadic life-styles make pursuing a career worthy of the name somewhat difficult. The problems raised by the conflicting requirements of dual-career couples, especially couples in which the husband is liable for duty 24 hours a day, are handled through implicit negotiation between husband and wife; talking of their future plans, young service members now use "we" more frequently than "I."

Wives also show a growing reluctance to participate in the life of the military community, and this reluctance extends beyond wives who work off base. Apart from work outside, the reasons most frequently stated for this rejection of the traditional military wife's life-style are that they resent the reproduction of the military hierarchy in their relationships with other military wives ("Mrs. Captain," "Mrs. Sergeant"); that they resent the informal social control that usually results from the traditional clustering of military families in the same housing areas or blocks of apartments; and that they sometimes regard on-base social life as a potential danger to their marriages. In the context of such a codified society as the military, relationships between men and women other than the purely conventional are apt to assume an ambiguous character and may lead to social and marital chaos.

These trends are slightly more marked among NCOs' wives than among officers' wives, probably because many officers' wives are officers' daughters, and therefore are more thoroughly socialized and less reluctant to assume informal institutional roles.

The net result of this evolution is that the traditional integrative function of military wives is being weakened. A substitute has yet to be found, and a new balance to be achieved.

In this regard, a useful hypothesis (as yet untested) is that the declining tolerance of informal social control, as it extends to private and family life among married service members, and the growing desire to live in the civilian sector away from other military families, is partly or wholly due to the influence of these "new model wives."

Societal Regard

The public image of the armed services is good; polls conducted on behalf of the Ministry of Defense show that about two-thirds of those questioned have a favorable opinion of the military and regard it as an effective organization. This finding must be viewed in the context of the societal landscape: in an age of reconstruction, after rapid social change in the Sixties and Seventies considerably weakened old norms, other major institutions or professions, such as the educational and justice systems, and even the medical profession, are all but a shambles, whereas the military apparently recovered quickly from its morale crisis.

Although the deaths of the paratroopers in Beirut in 1983 briefly revived old feelings and attitudes appreciative of the moral worth and grandeur of self-sacrifice, I have the impression (data are lacking) that the societal regard enjoyed by the military today is at least partly "occupational." It is based to a considerable extent on the prestige derived from a public image up of an effective mixture of high technology, travel, outdoor action, and—most of all—high pay for careerists and volunteers. Further prestige, especially as regards draftees, may be ascribed to the decline of antimilitarism and the concomitant rise of patriotism.[12]

Reference Groups

In the army, the primacy of the soldier's role over specialist's roles and the comparatively low rates of skill transferability inhibit horizontal identification. The temptation to identify with others who do the same type of jobs, with the same skills and about the same pay, is stronger in the navy and the air force. Even so—as repeated informal observations have revealed—it does not seem to extend far beyond the public sector, including civil service, nationalized utilities, and state-controlled civil and military aerospace industries. As regards the "services in blue," the significant division appears to be not so much between the civilian and the military as between the public and the private sector, as if a public service ethic were a natural outer limit to the shift away from the military institutional model and a powerful barrier against stronger occupational orientations.

Even identification with public-sector employees cannot be complete, however. Public-sector companies and, to a lesser extent, the civil service are strongly unionized, although the unionism does not have the same meaning as in the private sector. This tendency has always served as a foil to the services; the temptation to unionize has been entirely alien to the French military and in all likelihood will remain so. Unions are forbidden by law, and even the national union confederations, though urged on by extreme left-wing activists in the Seventies, declined even to try to introduce union cells in the armed services. Likewise in 1981, the new Socialist minister of defense, Charles Hernu, went on television within hours of his appointment to announce that contrary to rumors, and to what some had hoped, there would be no "soldiers', airmen's and sailors' committees" in the French forces. The determination of work conditions is therefore unilateral. This is not to say that in broader terms, implicit negotiation never takes place between individuals or groups in segments of the organization, which can be analyzed in terms of individual or collective power strategies. Still, open collective bargaining is abhorrent to the French military, to the extent that even the fairly paternalistic official system of representation and consultation sometimes produces uneasy feelings; the fear of creating a parallel hierarchy is always present.

Ministry of Defense and Contract Civilians

The very slight rise in the number of civilian employees is hardly significant, although it is comparable to decreases in the number of military personnel. Qualitatively, the overwhelming majority of civilians, apart from a few college professors detached to military schools and limited numbers of high-level administrators and experts, are in low-status categories; most fill blue-collar general support and maintenance of ancillary positions, which draftees cannot fill. The demarcation between civilian and military duties is sharply delineated, and even personal contact is limited. Therefore, little friction or sense of relative deprivation has developed. When the 39-hour workweek was adopted by Parliament, the measure, which had the effect of shortening the workday by one hour on Fridays, was extended after a time to service members. One suspects, however, that it was extended less as a result of pressure from the military than because civilian workers and service members often use the same (military) means of collective transportation to and from work. The Ministry of Defense uses the services of contract civilians, but military duties, except for the occasional "requisition" of long-range transport aircraft during overseas emergencies, are contracted out to civilian firms only on extremely rare occasions.

Legal System

In keeping with a presidential campaign pledge by François Mitterrand, military courts, which had been the target of bitter attacks in the Seventies, were abolished by the Socialist government in 1982. As this move followed on the heels of various restrictions of superiors' right to punish (under successive previous governments), it gave rise to serious misgivings and apprehensions among officers and NCOs, mainly in the army. In fact, so far it has had surprisingly little impact on the day-to-day operation of military organization. One can even argue that it has served to strengthen the enforcement of discipline: the old military courts had learned to treat draft violators and other such cases fairly leniently, probably to avoid unpleasant publicity and political trouble in the 1970s, when antimilitarist sentiment ran fairly high, whereas they were usually sterner toward "civilian" offenses, such as theft. It is much too early to venture definitive judgments now (established case law is lacking, and precedents need to be harmonized by the higher courts), but the trend paradoxically seems to be toward increased severity of civilian courts regarding "military" offenses, thus leading to the unanticipated consequence that civilian judges are "better" enforcers of draft law and disciplinary regulations than the military judges of old.

Separation and Discharge

With reference to separation policy, in the late seventies, enlistment contracts, though formally different in legal terms from private contracts, were brought into line with them in practice, at least in regard to individual enlistees. Although the organization rarely fired its members except for dereliction of duty, members could resign relatively easily. This policy led to excesses, however; private- or public-sector companies or even other public service organizations, notably police forces and the gendarmerie, tried—in most cases successfully—to entice young, fully trained NCOs away from the service into which they had enlisted before that service had had time to recoup training costs. Since then, resigning from the career force without showing good cause has become more difficult.

Military good-conduct certificates long ago lost their social and economic value on the outside. Indeed, a bad service discharge is no more of a handicap in the labor market than a similar discharge from a private company, and for draftees it is no handicap at all. The return of such certificates to common use would be a sure sign that the patriotic trend and the institutional orientations in civilian life are much stronger than was perceived until now, but we have not reached that point.

Post-Service Status

Veterans' benefits have traditionally been reserved for war veterans. These include pensions other than retirement or disability pensions, reserved jobs in the civil service, and special medical attention (although nothing resembling the GI Bill), hence the long controversy fueled by veterans' associations over whether the Algerian conflict was a real war or simply, as successive governments maintained, a long series of law-and-order maintenance operations.

Today, nonwar service veterans still have the option of entering the civil service upon leaving the military but with an important difference between NCOs and officers. The former are allowed to draw their civil service salary and (for those with 15 years' service or more) their immediate military retirement pension, mainly because given their seniority, that pension is usually only 35 to 40 percent of their last basic compensation. Officers, however, cannot receive their military retirement pension on top of their civil service salary until they have reached the upper age limit of their last military grade. (No such restriction exists in the private sector.)

Recently, in the context of unprecedented unemployment rates, the government has been under strong pressure to do away with or at least to restrict the right of retired NCOs and officers to a second career, but because age limits are lower in the military than elsewhere for functional reasons, such a move would amount to denying ex-service personnel their freedom of employment, a basic citizen's right enshrined in the Constitution. So far, the Ministry of Defense has successfully resisted such pressure.

Evaluation of Performance

The more capital-intensive a service or a branch thereof, the more segmented and quantitative the mode of performance evaluation. In that sense, the navy and the air force are certainly more occupational than the army, though that service is expected to catch up fast. Yet within the military framework, as in many parts of the civil service, this practice has obvious limits. Although the problem at the lower echelons is often to find the best combination of factors to achieve the norms set by the higher command, the higher echelons do not have a quantifiable final output by which to measure overall performance in peace (how much security did the armed services produce last year?) or sometimes even in war. The urge to quantify combat performance is apt to lead to absurdities; witness enemy body counts in Algeria or Vietnam.

The Reserves

Finally, a word or two on reservists is in order. Thanks to the universal egalitarian draft, the pool of reserve personnel is fed regularly by those who have completed their 12-month obligatory active-duty tour. Because France has opted for a small reserve force, rapidly mobilized and deployed, numbers are not a problem: the positions to be filled under the mobilization plan are oversubscribed three or four times. Therefore, quality is not a problem; with such numbers to choose from, selection on the basis of past performance, motivation, and distance between residence and location of mobilization units makes for a potentially high-quality reserve force. The problem, however, is a lack of money to make the paper reserve units real entities; the cadre and the equipment exists, but the outfits themselves rarely assemble for more than a few days a year. The reserves have not been a top priority in the last 20 years or so, but upgrading the reserve components, together with the advent of a "total armed force" concept, will probably be the next step in the modernization of conventional forces.

Reserve officers are generally highly institutional on most indicators applicable to them. Reserve NCOs and other ranks are very close to active-duty conscript NCOs and other ranks in attitudes, behavior, demographic variables, and most I/O indicators. Reserve personnel *were* active-duty conscripts three or four years earlier, serving in the same or similar ("twin") units or positions.

CONCLUSION

In summary, the French armed services have passed from the doldrums of the midSeventies to the prevailing high spirits of the midEighties, which the present chapter has tried to outline. The pay of career and volunteer personnel has improved substantially, and their prestige has risen. Recruitment and retention are satisfactory, and quality has risen dramatically. Draftees who agitated for better living conditions 12 years ago now uphold the very values that the military has traditionally stood for. Force levels have decreased, especially in the army, in keeping with the prediction of the "decline of the mass armed force." Technological modernization of conventional forces and new conventional doctrines have brought a new sense of purpose, notably in the ground forces.

This change owes much to the action of successive governments under both the modern conservative presidency of Valéry Giscard d'Estaing and the pragmatic socialist presidency of François Mitterrand. It probably owes more, however, to an exceptionally favorable context from the military standpoint—sharply rising unemployment and the gradual return of conservative and patriotic values.

The examination of I/O indicators has produced evidence that the French military, though less institutional today than it used to be, is still closer to the institutional than to the occupational end of the continuum. This general statement must be qualified by taking into account the variations of I/O modalities along with distinctions between services, personnel categories (officers, NCOs, other ranks; draftees and volunteers), and broad unit types (combat versus service support). Table 12-1 attempts to summarize the evidence gathered in the course of writing this chapter.

As a scientific instrument, Table 12-1 probably does not rank very high: imprecisely and often subjectively, it synthesizes bits of evidence that include established hard facts, the results of more or less fragmentary studies of varying scientific status, informal observations, and my impressionistic assessments and insights. Yet, for all its faults, if properly regarded as a rough approximation, it may clarify a number of points and facilitate comparisons, both internal and with other militaries. In particular, it will enable the reader to combine more conveniently the information provided in different columns (e.g., draftees in army support units). The bottom line of Table 12-1 supplies, in highly concentrated form, answers in the French context to some of the questions raised by the I/O model.

The navy and the air force are more occupational than the army; the latter appears to be making up for lost time in modernizing, but they can hardly be called occupational in absolute terms. Rather, they occupy the center of the continuum, mainly because, as this chapter has suggested, they have been protected until now from pure occupational orientations by a strong and still prestigious public service tradition. These two services are more civilianized than the army, though this status does not mean that they are militarily less effective. Likewise, officers and draftees—all things being equal—are more markedly institutional than NCOs and volunteers, who cannot be said to be occupational in absolute terms; in the officers' case, however, some of the institutional characteristics are declining. Finally, as might be expected, combat units are much more institutional than service-support outfits.

No attempt will be made here to predict the near future, but medium-term trends are a better bet. The technological upgrading of the army will probably make it slightly more occupational than it is today. The likely reduction of the ordinary draft tour to six or nine months instead of the present 12 months, and the correlative increase in the number of draft-motivated voluntary options and incentives, will result five to 10 years from now in a further accentuation of occupational tendencies in all three services, but most of all in the army, where draftees today still represent almost two-thirds of the total strength. Last, but by no means least, a marked improvement in the economic situation, notably employment, could wipe out some of the benefits enjoyed by the military in the recent period in regards to recruitment, retention, and quality, and could considerably accelerate "progress" toward the occupational end of the continuum. I feel, however, that the strong public-service tradition will prevent the armed services from ever becoming entirely occupational.

TABLE 12-1. Institutional/Occupational Variables and Trends in the French Armed Forces

	Army	Navy	Air Force	Officers	Career NCOs	Volunteers	Draftees	Combat Units	Support Units
Role commitment	III / II	I / O	I / O	II	I I (Army) / I O (Air Force)	II Army / OO Army Air Force	I / =	II / I	O / O
Basis of compensation	II	I	I	II	I	II	II		
Mode of compensation	OO	OO	OO	OO	O	=	III		
Range of compensation	III	II	II	II	III	II	III		
Residence	II	I	I	I (Junior/Bachelor) / O (Senior/Married)	I (Junior/Bachelor) / O (Senior/Married)	II	III	II	I
Spouse	I	I	I	I	I	=		I	=
Societal regard	I	=	=	=	O	OO	I	II	O
Reference groups	III	=	=	II	=	=		II	O
Legal system	=	=	=	=	=	=	=	=	=
Separation and discharge	I	I	I	II	I	=	=		
Postservice status	I	I	I	II	I	OO	OO		
General impression	III	=	=	II	I	=	II	III	O

III/OOO extremely institutional/occupational
II/OO markedly institutional/occupational
I/O slightly more institutional than occupational/occupational than institutional
= neutral, contradictory, or unclear

NOTES

1. Cf. Lucien Poirier, *Des strategies nucleaires* (Paris: Hachette, 1977).
2. Conscription was resumed in 1872 after a break in the "nation-in-arms" tradition, but this first draft law was neither very universal nor very egalitarian. In 1889, legislative steps were taken to achieve equality in military service, but it was the third draft law (1905) that instituted the fully universal egalitarian draft.
3. Michael Martin, *Warriors to Managers: The French Military Establishment Since 1745* (Chapel Hill: University of North Carolina Press, 1981).
4. For a useful clarification of this and related conceptual and doctrinal issues, see Lucien Parier, "La Greffe," *Revenu Defense Nationale* (April 1983). General Poirier's relatively short article does full justice to the complexity of the problems raised, which is far more than I could hope to achieve in these introductory remarks. See also David S. Yost, *Adelphi Papers*, no. 194–95 (London: I.I.S.S., 1985).
5. The gendarmerie is a special case. Most of its personnel are military (77,000 career service people, 9,000 conscripts, and only 1,000 civilians), and the gendarmerie would assume important responsibilities upon mobilization (territorial defense, general military police, and internal security). Its peacetime activities, however, revolve mainly around police and administrative jobs on behalf of the Ministry of the Interior, the Ministry of Justice, and other civilian ministries. Therefore, if the comparative referent is the U.S. military, the gendarmes are best left out of account.
6. In the assessment of trends observed recently, 1977 has been selected as the baseline year for comparison because Michael Martin's authoritative book *Warriors to Managers*, to which the reader is referred for earlier developments, covers the period 1945–76.
7. The amount of female participation is limited (3 percent overall), mainly because as long as conscription is in force there is no real need to tap that abundant source of personnel to compensate for quantitative or qualitative shortages in male recruitment. Numbers have risen only very slowly in the last 25 years. Women are distributed unevenly among the forces, with the largest concentrations in the air force and among NCOs and the smallest in the navy and among officers. Progress toward equality of status has likewise been slow, but, in the last eight years, significant breakthroughs have been achieved: gender segregation in training, promotion, and (with some restrictions) assignments has been eliminated, and equality of treatment is now the rule rather than the exception, though combat roles are still closed to women.
8. Pierre Saint Magary, "Vivre l'arme au pied," in Henry Mendras, ed., *La sagesse et le Desordre: France, 1980* (Paris: Gallimard, 1980).
9. Jean-Claude Roqueplo, "Les militaires: agents publics," in *Les Hommes de la Defense*, Cahier des Sept Epees, no. 21 (Paris: F.E.D.N., 1981).
10. This is mainly because in the last 30 or 40 years, union pressure has been much stronger at the bottom than at the top of the civil service hierarchy. In addition, the upper-class liberal tradition of disinterestedness has survived in the higher civil service, even though the social composition of the service has changed since the early days of the Third Republic.
11. Chantal Laharanne, "The French Army NCO's Wife: From Dependence to Interdependence through Negotiation," *Military and Society: The European Experience* (SO.WI International Forum, 4, and Centre William I. Thomas, IEP de Toulouse, 1984).
12. Polls conducted regularly on behalf of the Ministry of Defense (SIRPA) show that the percentage of those "with no anti-militaristic feeling at all" rose from 52 to 62 percent between 1976 and 1983. Likewise, "national symbols" are felt to be important by 70 percent (as opposed to 64 percent in 1976).

XIII

Australia

NICHOLAS A. JANS

Australia, with a population of almost $15\frac{1}{2}$ million, has a defense force strength of 98,000, including permanent forces and active reserves. This represents a proportionately small military among Western parliamentary democracies. About half of the Australian Defense Force (ADF) is found in the army; the remainder is more or less equally divided between the air force and the navy.

The bulk of the ADF is based in Australia. The air force has an operational base in Malaysia; apart from small contingents with the United Nations, this is the only Australian military presence overseas. Yet for almost three decades after the end of World War II, Australian forces were frequently engaged in active service in Asia: first in Korea, then in Malaya and Borneo and most recently in South Vietnam. In these latter two campaigns, the army gained its high professional reputation as a counterinsurgency force. Australian strategic policy has changed, however, since the early 1970s: between 1945 and the end of Australian involvement in Vietnam in 1973, it was essentially a policy of forward defense, but now it is focused on the defense of continental Australia. This change has required fundamental rethinking of doctrine in all three services, a process that is currently far from complete.

The importance of the military in Australian history is symbolized by Anzac Day, one of Australia's national public holidays. Anzac Day tends to be to Australia what Bastille Day is to France. The Australian and New Zealand Army Corps (ANZAC) first fought at Gallipoli and went on to become an elite formation in the British army in France. The Anzac image — an image that had an unassessable but possibly significant effect on the national identity—is based on the Anzac's twin reputations for dash and courage in action and for individualism and what some saw as poor discipline (the Anzacs frequently refused to salute British officers, for example). However, the two sides of the Anzac character are mutually supportive, and this character still tends to influence the attitudes of modern Australian military personnel. They tend to be

TM –O

patriotic, but without ideological commitment, and loyal, but more to the primary group than to the service as a whole. Individual military values may have some of their roots in this Anzac image.

The ADF is presently an all-volunteer force, although selective conscription for the army was used during the Vietnam War between 1966 and 1972. Most officers hold a permanent commission, but some hold a short-service commission. A permanent commission, in principle, obligates the recipient to indefinite service, but in practice, officers' resignations are usually accepted unless a return of service for service training is required.

- **Navy:** General list (GL) officers receive common cadet training at the Royal Australian Naval College at Jervis Bay. GL officers constitute the seamen, engineering, and supply branches.
- **Air Force:** Air force officers are trained at a variety of schools, depending on the branch. The Royal Australian Air Force (RAAF) Academy at Point Cook (four-year course) is the entry point for general duties branch officers who wish to gain a university degree before commissioning, but these officers are a minority in their branch.
- **Army:** General service officers are trained as cadets at either the Royal Military College at Duntroon (four years) or at the Officer Cadet School at Portsea (one year). After 1986, all cadet officer military training will be centralized at Duntroon.

There has been a marked trend toward university studies for officer cadets in all three services. All Duntroon and Point Cook graduates are already required to complete a degree before commissioning or employment. In 1986, tertiary studies for all ADF cadets (including navy) will be concentrated at the Australian Defense Force Academy at Canberra.

Among other ranks, sailors, soldiers, and airforce personnel are enlisted for terms ranging from three years (army only) to nine years, and they may reengage or be discharged after this initial period. The variety of employment is large, and the more technical employments require the longer engagement periods.

INSTITUTIONAL AND OCCUPATIONAL CONCEPTS

The institutional-occupational (I/O) thesis, as originally formulated by Charles Moskos, assumes a continuum ranging from a military organization highly differentiated from civilian society to one highly correspondent to civilian structures. The idea of a continuum implies a unidimensional concept of I/O orientations; that is, a person who has a strong institutional orientation will have a weak occupational orientation, and vice versa.

In light of the wide interest in the I/O concept, the issue of operationalization of the dimensionality of this concept has not been explored by researchers as thoroughly as might be expected. To explore this aspect further, I constructed questionnaire items in collaboration with J. Graham Pratt to measure I/O orientations. These items were used in a large-scale 1984 survey of ADF officers and a 1985 survey of noncommissioned officers.[1] The scale items are given in the appendix later in this volume. These survey data, as well as descriptive analyses, will be used to ascertain institutional and occupational trends in the contemporary Australian military. We will treat in turn such key I/O issues as the societal legitimacy of military service, reference groups, the basis and mode of compensation, housing, and contemporary trends in spouse relation and educational levels.

Legitimacy

If I/O trends in the ADF could be summarized in one sentence, it would be as follows: the legitimization of military service remains strongly normative, but with simultaneous trends toward a more industrial attitude by service personnel.

Normative Values

The two norms that (as Charles Cotton argues) represent the military ethic are that service members must do their duty regardless of personal and family consequences (primacy) and that they are on duty 24 hours a day (scope).[2] Acceptance of such norms is an acceptance of the legitimacy and pervasiveness of military discipline: to accept such norms is to take on the military self-image in the fullest sense. Australians who join the ADF can have little doubt of the presence of military discipline and of the existence of these norms; thus, to some extent, they embrace them voluntarily by joining.

The public in Australia has only a vague concept of what the military is really like.[3] In contrast to the United Kingdom and United States, films, TV shows, and news stories about the services are rare. The public becomes aware of the service only in times of emergency, such as when RAAF aircraft air-drop supplies as part of flood relief or when soldiers assist in fighting bush fires. These glimpses reinforce the residual image from the wars in which Australia has participated: a tough, disciplined, professional fighting force, emphasizing the norms of the military ethos—the Anzac image that was sketched earlier.

Because enlistment is voluntary, enlistees are assumed to accept the military ethos as legitimate.

Acceptance of Military Associations

Regardless of these norms, military associations in the ADF have been increasingly acceptable, as evidenced by the results of two research projects. In his research on army officers in the late 1970s, Pratt found that 30 to 40

Nicholas A. Jans

TABLE 13-1. Support for Military Associations by Service

Preference for type of collectivism[a]	Officers[b]				NCOs[b]		
	% Navy (N = 380)	% Army (N = 460)	% Air Force (N = 454)	Mean of Officers	% Army (N = 75)	% Air Force (N = 115)	Mean of NCOs
No kind of collective representation	2	2	2	2	5	2	3
A representative body to consult with and advise the government on pay and conditions	29	28	24	27	42	44	43
A representative body to negotiate with the government	51	55	56	54	45	42	44
A representative body to negotiate and prepare, if necessary, to take mild industrial action	14	11	14	13	7	8	7
A representative body to negotiate which is prepared to take full industrial action	4	4	4	4	1	4	3

a. In response to the question: "In terms of representation for members of the Defense Force on pay and conditions, which ONE of the following options would you prefer?"
b. Service differences are not significant.

percent would accept in principle some form of mild collectivism, such as an association prepared to negotiate with the government but forbidden to strike.[4] Five years later, the proportion of army officers with these views was close to 70 percent, as Table 13-1 shows.

This trend could be attributed to three factors: the changing national industrial relations climate, where professional groups who were once loath to strike are no longer reluctant to do so (e.g., teachers, police, firefighters, doctors); a civilianization of the industrial relations practices and procedures in the ADF; and what many service personnel see as an apparently progressive erosion of ADF conditions of service, with service chiefs seemingly unable to protect the individual's interests.

The first of these trends tends to legitimize the notion of collectivism for service personnel who might previously have regarded it with suspicion. The second trend also tends to legitimize civilian-style industrial relations practices in the ADF. The third tends to create a feeling that the chain of command can no longer be relied upon to protect individual interests, and that collectivism is the only workable response if interests are to be protected.

Civilianization of Military Industrial Relations

Before the late 1960s, the military industrial relations system[5] generally displayed the characteristics of a traditional model: terms and conditions were set exclusively by the employer (the Australian government); collective organization and action were forbidden; grievances were redressed on an individual basis through the chain of command; rules, pay, and conditions reflected the distinctive features of the military life-style. Resigning was difficult for officers, for example, and officers who resigned did not qualify for pensions. In addition, a large number of tax-free allowances were available to service personnel, including a marriage allowance.

During the 1970s, some important changes were made in the Australian military industrial relations system.

1. As a result of an inquiry into remuneration, a civilian-style salary structure was adopted.[6] Marriage and other allowances were removed, together with the "daily rate" concept.
2. A quasi-arbitrational independent body—the Committee of Reference for Defense Force Pay—was established to report on aspects of ADF remuneration.
3. Regulations associated with individual rights and the expression of grievances were liberalized. Resigning became much easier, and grievances could be directed to a defense force ombudsman and other forms of administrative appeal.
4. A new retirement scheme was established, providing a pension and a commuted lump-sum payment after a minimum of 20 years of service.
5. The minister of defense in the new Labor government of 1972 publicly raised the issue of servicepersonnel's associations.[7]
6. ADF industrial relations were centralized after the amalgamation of the three service headquarters into a single Department of Defense headed by a public servant (the secretary). A complex network of committees was set up in the Department of Defense to determine industrial relations in the services, each comprising a mixture of service officers and public servants. In essence, this network is under the sponsorship of the secretary of the Department of Defense, not the chief of the defense force. Thus, the development of conditions of service policies is now not solely the responsibility of the chiefs of the defense force.

These changes affect not only the way industrial relations are practiced in the ADF, but also the way service personnel think and feel about industrial relations. The attitudinal consequences of the changes include

- an increased general awareness of and interest in the methods by which Australian Public Service pay rates are set, including the role of employee associations (the APS is heavily unionized);
- a lessened commitment to a long-term military career, with the RAAF in particular becoming what air force personnel managers call "a 20-year air force";

- a greater willingness to seek redress of grievances outside the chain of command (in some cases the chain of command is bypassed completely); and
- an increased awareness and heightened specialized skill among those service officers employed in the staff branches established to manage conditions of service issues.

In the 1980s, the defense industrial system has moved even further from the old paternalistic practice with the establishment of a Defense Force Remuneration Tribunal, which has the legal power to make determinations that can be disallowed by Parliament alone.

Thus, in the 1980s, Australian services' industrial relations systems are tending both to raise the sensitivity of individual service personnel to industrial relations matters and to separate the interests of top management from those of the rank and file on such matters.

Erosion of Conditions of Service

A number of incidents during the 1980s have indicated to service personnel that conditions of service are being eroded, despite the intervention of their service chiefs. ADF officers seem to have developed the belief, which they voiced frequently during my interview survey of 1984, that "the system" was no longer capable of protecting the individual service member's interests.

This belief almost certainly contributed to the formation of the Armed Forces Federation of Australia (AFFA) in late 1984. In the past, certain legal decisions on national industrial relations policies had placed practical difficulties in the path of any association of service personnel, but more recent decisions made such an association more viable.

How many persons have officially joined the AFFA is not yet known, but the figures in Table 13-1 suggest that a large "market" of potential members is available.

The changes in the services' industrial relations climate are a plausible determinant of an increase in occupationalism. Because these changes have been concerned essentially with conditions of service, the trend toward occupationalism seems to have had little effect on an acceptance of the military ethos. The modern service member appears to be quite capable of saying, "I put the service before myself," and, almost in the same breath, "If we leave conditions of service to 'the system,' there'll soon be nothing left." This attitude is summed up by an officer who wrote at the end of his questionnaire in my study:

> I am amazed and incredulous about my answers as I used to feel strongly that unionism and the military don't mix. However, the Services and the Serviceman has [sic] had so little say and been given so little attention that only strong action appears to get the message across!

Role Commitments

A belief that the service member's job specification is open-ended—that he or she is required to perform any lawful task the service demands—is consistent with high institutionalism. The extent to which such tasks are actually required of service personnel will reinforce or weaken institutional orientations, as applicable. Moreover, occupationalism may be increased by a weakening of the beliefs that role commitment is diffuse and that a service member is on duty 24 hours a day, if such beliefs are associated with other issues which pertain to conditions of service. For example, adherence to standard working hours may encourage the idea of compensation for nonstandard hours worked (i.e., overtime). This compensation may not be monetary; in the 1970s, common practice became for service personnel to be granted "days in lieu" for weekend time spent on exercises or other duty.

Another factor is career specialization. As a matter of policy, sailors, soldiers, and air force personnel are highly specialized in their employment. For officers, the policy is more flexible. General service officers in the army, general list officers in the navy and general duties officers in the air force are all expected to follow a varied path throughout their careers. Officers in other employment categories (navy and air force engineers and supply officers, for example) are more specialized, but are still expected to fill nonspecialist posts if necessary. Such diversity strengthens institutionalism by continually reinforcing the principle that service needs have priority over the officer's personal career needs (see also the discussion on career training).

Among the first type of officers—the generalists—practices and attitudes toward generalization still vary. My research in 1984 showed that many navy sailors and air force pilots follow highly specialized careers in their first decade of service, in contrast with the great majority of army officers, whose experience is much more general. This is especially true of pilots, whose training is so expensive that their service is reluctant to employ them away from flying. Such specialization tends to reinforce pilots' identification with flying as their career rather than with the air force (see also the discussion on reference groups) and tends to make many pilots regard their service career as limited by the stage at which they can no longer fly regularly. My interviews revealed that many pilots actively seek flying jobs outside the air force after about 10 years of service.

Reference Groups

Army officers in the mainstream of their service (those who are specialists in military operations) tend not to identify with reference groups outside their service. In a 1979 study, for example, army officers were asked to complete the Vocational Preference Inventory (VPI), which is based on the assumption that because people have stereotypes of many occupations, they can judge the extent to which they would like or dislike each of 160 listed jobs.[8] In the pattern of their expressions of attraction or repulsion, they usually reveal the bases of their self-images. The majority of this army sample could identify positively with only a handful of civilian occupations and scored well below the norms for the population at large on all the dimensions of the VPI.

Conversely, air force technicians have a strong sense of identification with civilian occupations. In a 1978 study, when asked to choose words from a list to describe themselves, many technicians selected civilian descriptors in preference to military descriptors.[9] The choices, however, differed among the types of technician. The more similar the employment was to a civilian employment, the more likely the technician was to choose a descriptor such as "tradesman" or "manager" over a term such as "airman" or "serviceman." Thus, civilian descriptors were chosen by 39 percent of armament technicians, by 46 percent of both aircraft technicians and electrical and instrument technicians, and by 60 percent of radio technicians.

Air force pilots often identify with the occupation of flying rather than with the profession of arms. Among my interview sample (N = 129) across many ranks in all three services, the pilots were the only subgroup who consistently were very clear about why they joined their service: they joined to fly. Many of them would no doubt agree with the young flying officer who said during an interview in the study, "The Air Force tells me I'm an Air Force officer first and a pilot second; I say I'm a pilot first and a.pilot second." Attitude measures tend to support this: pilots who are currently posted to flying jobs tend to have the highest job involvement of any ADF officer in a professional category, but pilots who are not in flying jobs tend to have the lowest.[10]

Basis and Mode of Compensation

Pay increases with rank, with the important exception that junior officers are paid less than senior sergeants, but the overriding factor in the ADF is the sharp compression of salaries. Thus, a major general earns only three times the salary of a beginning lieutenant, and a senior noncommissioned officer (a warrant officer in the ADF) earns 2.6 times the rate of an entering private. Within the other ranks, differences in pay are made according to skill level. Skill level is related largely, but not entirely, to time in grade.

In recent years, the ADF has moved toward a mode of compensation based more on direct salary and allowances than on noncash forms, but this is more a simplification of past practices than a significant change. Initial issue of uniforms is still free, as is medical and dental treatment for the service member, and a nontaxable uniform maintenance allowance is paid. A levy is made for the rations and quarters of those living in barracks, but this expense is subsidized so that the member pays only costs or less than costs.

In contrast to the American forces, there are no subsidized shopping facilities for service members and their families and no free medical and dental treatment for dependents or veterans.

Housing

Service personnel tend to live off base when they are married or in a de facto relationship. Many either do not have married quarters (MQs) available to them or have their own homes in the area in which they work. The former group is assisted with the rent of a local flat or house, but even those who occupy an MQ may live away from the military community because the MQs are often scattered through the suburbs of the adjacent city or town. In 1985, among those entitled to an MQ, only 10 percent of navy personnel, 7 percent of army personnel, and 17 percent of air force personnel occupied a MQ on base; the remainder lived in the civilian community.

The provision of housing for service personnel is currently under review by the government. A preliminary report of the review body has raised the possibility that the government may take less responsibility for defense housing than it has in the past and that such responsibility may be placed more on the individual.[11.] Reactions by service personnel and by the AFFA on their behalf indicate that this is another factor that has the potential for increasing occupationalism in the ADF.

CAREER TRAINING

Career training can affect institutionalism by the extent to which it reinforces or weakens the concept of service-oriented role models as opposed to employment-oriented role models. In this sense, career training is a variable that extends the discussion on reference groups. We propose that career courses in which members of different employments or professional categories come together for common training, focusing on service issues rather than on job issues, reinforce the service role model and weaken the employment role model. As a corollary, career training based largely on employment-specific courses may weaken the service role model and reinforce the employment role model, especially if the latter has an obvious civilian equivalent or basis.

The three services follow different practices in this regard. The army has a strong service orientation in its career training for officers up to the rank of lieutenant colonel and for all soldiers. The navy and the air force have much less common career training for both officers and enlisted people.

Army

Precommissioning training, whether at Duntroon or at Portsea, is a common course for all aspirants to the general service officer (GSO) division. After this, GSOs come together again for common courses at least three more times over a period of 15 years, except for the Command and Staff College course, which not all officers attend. In addition, many employment-specific courses are offered.

A similar pattern exists for soldiers. Except for apprentices, all soldiers attend a common recruit-training course. Promotions to corporal, sergeant, and warrant officer class 2 require attendance at a common course, often conducted at a central location. Again, there are also many employment-specific courses.

Navy and Air Force

Both services have a number of employment-specific courses for officers, but offer very few common career courses apart from staff college. Navy engineer branch officers, for example, attend no career courses after attaining their various certificates of competence as junior officers, unless they attend staff college. Air force officers attend common courses for a total of 10 to 13 weeks before employment as officers and an eight-week basic staff course at the rank of flight lieutenant. Air force officers also complete specific on-the-job training assignments under their work supervisors and a two-year correspondence course as a prerequisite for the advanced staff college course, but these activities are self-paced and do not involve attendance at formal courses.

Different patterns apply for sailors'and airforce personnel's career training in the navy and the air force. In the navy, promotion to leading seaman and petty officer requires attendance at a course that has standardized training objectives but is conducted by units. In the air force, promotion to sergeant is usually preceded by attendance at a management course at a central service school. Generally, however, NCOs in these services do not receive the kind of common career training that is given to their army counterparts.

CONTEMPORARY TRENDS

Spouse Relations

The 1984 study of ADF officers and their families showed that wives tend to be supportive of their husbands' careers. Wives who could be classified as "fully supportive" or "conditionally supportive" constituted 71 percent of those sampled, based on their scores on a scale of wife's support for her husband's career. Only 8 percent could be classified as "alienated" from or opposed to the husband's career, and the remaining 21 percent were "ambivalent" in regard to their involvement and support.

Whether or not this relatively large group of "ambivalents" represents a growing trend remains to be seen. A wife's "sex-role image"—her feminist values—has a negative relationship to her support for her husband's career, and because younger wives tend to have a less traditional sex-role image, in time, the average level of feminist values among officers' wives is likely to be more contemporary. If so, the level of support for the husband's career is also likely to be somewhat lower in the future, although sex-role image exerts a minor influence on support compared with the influence of other variables, such as quality of family life.

Officers and their spouses feel very strongly about the effects of the service life-

TABLE 13-2. Officers' Educational Level and Institutional and Occupational Orientation

Education Level	N	Mean Institutional	Mean Occupational
Higher degree (PhD/Master's)	47	4.23	4.19
Graduate diploma	83	4.24	4.00
Bachelor's degree/diploma	408	4.25	3.95
Partly completed bachelor's degree/ diploma	131	4.61	3.96
HSC/technical college certificate[a]	333	4.44	3.89
Less than HSC	108	4.58	3.83
	1,110	F = 4.50	F = 1.54
		p = .0002	n.s.

[a]HSC is "Higher School Certificate," certifying graduation from the senior level at high school.

style on children's education, so much so that the higher the child's level of schooling, the less willing the officer is to change locations even for career reasons. That is, officers with children in high school are less willing to move their families according to the imperatives of career management than those with children in primary school or not in school.

Another trend among service families is the incidence of wives in paid employment. Among officers' families, 53 percent of wives work in paid employment (10 percent greater than the national rate for married women), and they are more likely to do so if they are young or have tertiary qualifications (68 percent of wives aged 19 through 25 are in paid jobs, as are 70 percent of wives of all ages with tertiary qualifications). A wife's employment status does not seem to affect her support for her husband's career, but an officer is less willing to move if his wife is psychologically involved in her career.

As yet, no similar data are available for the wives of sailors, soldiers, or airmen.

The wives in the survey tended to support involvement in service social life. When presented with the statement "If I attend 'official' or mess functions with my husband it is only because I feel that not to do so would be harmful to his career," 74 percent disagreed, 18 percent agreed, and 8 percent were not sure, with no age group differences in responses.

Educational Levels

As noted earlier, there is a strong trend toward raising the educational standards of ADF officers. Table 13-2 shows that in the future this trend could have a depressant effect on levels of institutionalism in the officer corps. Officers who are educated at or beyond the university undergraduate level have lower levels of institutionalism than those whose formal levels of education are lower. On the other hand, educational level has no effect on occupationalism.

The I/O differences are statistically significant but are not large in practical terms. Tertiary-educated officers are still institutionally oriented, but apparently they are more likely to be critical of existing policies and norms. This tendency, however, does not make them more self-interested regarding pay and conditions of service, which is probably good news for the services; an inquiring and critical outlook is a worthy goal of tertiary education.

SUMMING UP

To sum up the many factors that affect I/O orientations, three elements—legitimacy, role commitments, and reference groups—stand out as particularly important. Although the military ethos seems still to be central to military service, there is a growing acceptance of the legitimacy of collective representation by service personnel. This trend seems to be the result of a perceived inability of the service hierarchy to control conditions-of-service issues and is manifested in the recent formation of the AFFA.

Role commitments and reference groups, as factors, are linked. By definition, increasing specialization limits the generalist philosophy of career development, and specialization tends to be most prevalent among those employments that have obvious reference groups outside the services. The most obvious examples are pilots and air force technicians, many of whom identify as much with civilian reference groups as with service role models, if not more. Career training policies seem to be important in this regard.

Wastage rates vary according to market forces. The wastage rate for armament technicians is much lower than for electronics technicians, for example. To maintain an establishment strength of around 600 armament fitters, the air force presently trains only 25 per year (4 percent), whereas they train 150 electronics technicians (8 percent) each year to maintain an establishment of about 1,800.

RAAF personnel managers refer to a "20-year air force," by which they mean that few enlisted personnel serve past the pension-qualification point of 20 years' service. The long-term average of those who do so is around 4 percent. Recently, during Australia's somewhat straitened economic situation, the rate has risen to around 6 percent, but some RAAF personnel managers believe that this figure will slip when the economy improves.

RAAF response to the high wastage rate among electronic tradespeople is career-oriented rather than economic. (Most tradespersonnel are already at the top of the salary range at each rank.) A military occupation called the *system technician* is to be developed to absorb electronic specialists at the senior NCO level. Such personnel have usually achieved a higher level of early training, taking them to "diploma of technology" (tertiary level) standard. After employment in this field as a senior NCO, such a technician may be offered a commission. The scheme is still in the early stages, so its success cannot yet be judged.

EFFECTS OF I/O ORIENTATION ON CAREER INVOLVEMENT

Career involvement is emotional engagement with a career role and the extent to which a person feels a psychological affinity with that role.[12] We measured career involvement along with a number of variables, including I/O measures.

TABLE 13-3. Officers' Career Involvement: Effect of Institutional
and Occupational Orientations
Frequency Distribution

Institutional Orientation	Career Involvement				
	Low (N = 234)	Moderate (N = 318)	High (N = 548)	Totals	Chi Square
Low	41	17	6	16	205.16
Moderate	36	43	29	34	$p = .0000$
High	23	41	66	49	
Occupational Orientation					
Low	14	26	42	31	124.52
Moderate	41	51	46	46	$p = .0000$
High	45	23	12	22	

Table 13-3 shows that officers with high institutional orientations and low occupationalism tend to be highly involved in their careers, and those with low institutionalism and high occupational orientations tend to be uninvolved. Similar relationships were found for noncommissioned officers. When statistical controls were introduced, institutionalism had a greater effect on career involvement than occupationalism on career noninvolvement.

To sum up the effects of I/O orientations on career involvement, each orientation is likely to exert a different type of influence. As noted, institutionalism is an indicator of the person's self-image and affinity with the military work style. People with high occupationalism are influenced more by economic considerations and are less personally involved in their careers because their futures are less bound up in the organization.

PROFESSIONAL CATEGORY

Given the notion of I/O orientations as an indicator of self-image and the socializing effect that the work environment has on self-image, we might imagine that institutionalism and occupationalism would vary across professional categories; that is, we might expect high institutional orientation and low occupational orientation among those whose primary professional role is combat, and the opposite pattern among those whose employment is far removed from the battlefield.[13]

Table 13-4 shows data that test this hypothesis.

Officers

In only one case—army officers' institutionalism scores—is this hypothesis supported, and even then the level of significance is not high. Two-thirds of

TABLE 13-4. Institutional and Occupational Orientation by Professional Category

Professional Category	Institutional			Occupational			Chi-square		
	Low	Moderate	High	Low	Moderate	High	N	IO	OO
Navy Officers									
Seamen	16	36	48	34	40	26	155	n.s.	n.s.
Engr/supply	18	42	40	27	54	19	100		
Specialist	21	39	39	13	26	16	55		
Total	18	39	44	30	46	22	310		
Army Officers									
Combat arms	10	23	67	39	43	19	124	13.5	n.s.
Combat support	16	31	52	32	48	20	86	*p* = .01	
Logistic support	10	28	62	36	48	16	126		
Specialist	18	41	41	29	38	33	56		
Total	13	29	58	35	45	20	392		
Air Force Officers									
General Duties	28	34	38	32	46	22	117	n.s.	n.s.
Engr/supply	17	37	47	25	47	28	174		
Specialist duties	10	42	48	30	57	13	60		
Specialist	20	37	44	29	47	24	45		
Total	19	37	44	28	48	23	396		
Army NCOs									
Infantry	50	29	21	25	43	28	32	n.s.	n.s.
E&M engrs	38	40	22	23	52	25	45		
Air Force NCOs									
Nontechnical	14	36	50	23	43	34	36	17.7	n.s.
Technical	49	34	17	18	48	34	80	*p* = .001	

Figures in institutional and occupational columns are percentages.
Percentages may not total 100 because of rounding-off.

officers in the combat arms (armor, artillery, and infantry) have high institutional scores, but this proportion is not much higher than for officers in logistic support corps. Both professional categories have a greater proportion of officers with a strong institutional orientation than we find among officers in combat support arms (engineers, aviation, and signals). The relatively low institutional scores of this latter group are puzzling.

I/O scores do not vary by officers' professional category in the other two services, nor do they vary for the sample as a whole when it is broken down by professional category (this latter analysis is not included). In the navy, the pattern of I/O scores is in the expected direction, but the trend is not strong enough to be statistically significant. The differences within the air force are not significant, although they are consistent with earlier observations of air force pilots' attitudes and professional identification with the role of pilot rather than officer.

NCOs

The data presented here are taken from a pilot study of army and air force NCOs, which used a questionnaire with items similar to those on the officers' questionnaire. The institutional levels of infantry NCOs in this study are surprisingly low: only one soldier in five could be said to have a high institutional orientation, and no significant differences occur between the I/O levels of infantry NCOs and NCOs in the electrical and mechanical engineers. In the air force subgroups, nontechnical NCOs have significantly higher levels of institutionalism than their technical peers, who appear (predictably) to have the lowest institutional levels and the highest occupational levels of any samples of the professional categories.

The NCO pilot study shows the need for further research. The data suggest a substantial discrepancy between the military values of officers and those of NCOs in both the army and the air force. Further speculation on this phenomenon must await systematic study.

CONCLUSIONS

This chapter has proposed that institutionalism and occupationalism are distinct constructs, although they are related empirically and conceptually. I discussed the sociological factors in the ADF that contribute to I/O orientations and suggested that the major influences on these orientations are legitimacy, role commitment, and reference groups. Although the military ethos still seems to be central to military service—at least among officers—at all ranks, service personnel show a growing acceptance of the legitimacy of collective representation. This acceptance seems to be the result of a perceived inability of the service hierarchy to control conditions of service, and it is manifest in the recent formation of a service personnel's association.

Role commitments and reference groups are linked as factors. Increasing specialization, by definition, limits the generalist philosophy of career development, and specialization tends to be most prevalent among those military employments that have obvious reference groups outside the services. The most obvious examples are pilots and air force technicians; research on these groups has shown that many tend to identify with civilian rather than service reference groups.

I also addressed other issues, including specific information on the two last-mentioned professional categories, career training, and the effect of I/O orientations on career involvement, and I assessed the I/O levels in particular professional categories among ADF officers and NCOs. Although the data for the latter are drawn from a pilot study and thus are highly tentative, the apparent gap between officers' and NCOs' I/O levels would be a cause for concern if such levels were confirmed by further study. Such study would benefit from a consideration of the factors discussed earlier as the potential source of variance in institutional and occupational orientations.

NOTES

1. These research projects included a large-scale study of officers in the ADF and a pilot study of army and air force NCOs. The first study, conducted in 1984, involved a stratified random sample of

officers between the ranks of lieutenant and colonel (or equivalents) across all professional categories. Data were gathered through interviews with officers (N = 129), a questionnaire survey of officers (N = 1300), interviews with wives (N = 35), and a questionnaire survey of wives (N = 943). The results are forthcoming in Nicholas A Jans, *Careers in Conflict: Service Officers and Families in Peacetime* (Canberra: National Defense College, in press). The second project was a pilot study conducted in 1985, using a questionnaire similar to that of the officer study. The subjects were NCOs attending promotion courses at the School of Infantry, the Royal Australian Electrical and Mechanical Engineers (RAEME) Training Center, and the Management and Instructional Methods Squadron of the RAAF School of Technical Training. We hope to expand this second study into a project similar in aims and coverage to the first.

2. Charles Cotton, "Institutional and Occupational Values in Canada's Army," *Armed Forces & Society* 8 (Fall 1981): 99–110.

3. Australian National Opinion Polls, *Community Attitudes towards Australia's Defense Force: A National Communications Study* (Presented to the director general of recruiting, October 1980).

4. Graham J. Pratt, *Perceptions of Industrial Relations Among Army Officers* (Unpublished PhD thesis, University of New South Wales, 1982). Pratt's sample was 208, composed proportionally of officers ranking from second lieutenant to colonel, inclusive.

5. Much of the following is taken from Pratt, *Perceptions*, and Pratt, *An Australian Application of the Military Sociology Convergence Theme* (Paper presented at the Annual Conference of the Australian Study Group on Armed Forces and Society, Duntroon, June 24, 1983).

6. Australian Government, *Report on Committee of Inquiry into Financial Terms and Conditions of Service for Members of the Regular Armed Forces* (June 1971 to December 1972, Reports 1 to 5 and Final Report).

7. Lance Barnard, *Australian Defence: Policy and Programmes* (Melbourne: Victorian Fabian Society, 1969).

8. Nicholas A. Jans, *Work Involvement and Work Satisfaction: An Investigation of Two Indicators of Work Adjustment in Organizations* (Unpublished Ph.D. thesis, University of New South Wales, 1979); John L. Holland, *Manual for Vocational Preference Inventory* (Palo Alto, California: Consulting Psychologists Press, 1975).

9. Michael J. Rawlinson, *Report on Labor Turnover in the Technician and Equivalent Trades of the Royal Australian Air Force: An Economic Analysis.* (Report submitted under the Defense Fellowship Scheme, 1978.)

10. Job involvement of officers was measured by a six-item scale with an alpha coefficient value of .85. (Example: "The performance of my present job is a good test of my skill and ability.") In this context, *job* means the current appointment, not the career as a whole. Officers were divided into subsamples by professional category and by whether the officer was employed in an operational unit. As noted in the text, general duties branch officers had the highest job involvement of any subsample when they were in an operational unit, and the lowest when they were in another type of unit (respective mean scores, 5.20 and 3.88).

11. *Task Force on Australian Public Service and Defence Force Housing Programs: Interim Report of Program Effectiveness Review* (January 1985).

12. Example item: "I am very much personally involved in my career as a Navy/Army/Air Force officer." The career involvement scale had an alpha coefficient value of .81 (officers) and .70 (NCOs).

13. Cotton, in "Institutional and Occupational Values," did show a strong relationship between institutionalism and the combat arms in the Canadian forces.

XIV

The Netherlands

JAN S. VAN DER MEULEN

Charles Moskos's reflections on the armed forces have an unmistakably concerned undertone. Obviously he is worried that as a net result of a few changes the armed forces as an institution are heading the wrong way, not only in the United States but also in the rest of that conglomerate of countries that we loosely call "the Western world." In this chapter, I'll test Moskos's worries on the Dutch military, which is very much a part of that Western world. These worries, though evidently military, are somewhat a matter of personal taste and judgment. They are embedded in a theory that touches many things and people, not only the military. The inclusive character of the institutional-occupational (I/O) theory makes it attractive; theories that touch only one or two subjects are not worth the paper they are written on. This encompassing quality, however, also makes it necessary to state in advance how I will apply it to the Dutch case.

As I see it, the theory states essentially that the armed forces as an institution are losing sense, meaning, legitimacy, and, as a result, quality and effectiveness. The erosion of the underlying institutional values is symbolized, and perhaps caused, by developments on three interdependent levels: the military as a community, a subculture, a way of life; the military as a collection of professions, occupations, jobs; the conditions of employment in the military. At the first level, we deal with variables like "residence," "spouse," "sex roles," "legal systems"; at the second level with "role commitment," "evaluation of performance," and "reference groups"; at the third level with "pay" and "compensation." "More privacy," "more specialization," "more hard cash," would be a suggestive though some what oversimplified summary of trends on the three levels. *Civilianization* would include all three, as would *rationalization*. In a more abstract way, one could even detect a development from *Gemeinschaft* to *Gesellschaft*. This brings us back to the heart of the I/O theory: the fate of the values that give the institution its meaning.

The meaning and significance of the military is not determined by only the insiders. In reality, the legitimacy of the armed forces rest as much, or even more, in the hands of the outsiders, so we'll have to talk about armed forces and society. The I/O theory explicitly invites us to do so, and especially in this regard it is more than just another analysis of developments in the military profession. It also differs from that kind of analysis by not focusing exclusively on officers. Noncommissioned officers, corporals, privates, and sailors also count, which is rather unusual in military sociology.

In looking at the Dutch case, I will be eclectic in method and use heterogeneous data, which I am tempted to call "figures and other rumors." Those other rumors are made up partly of stories told by conscripts, sergeants, majors, and generals; partly of reports, plans, and policies that the Defense Department has in store; partly of the views of associations and unions; and partly of surveys conducted among the military and the nonmilitary. These surveys contain some of the figures I will use.

I will try to use this material in essay form to suggest the mentality, the culture, and the mood that apparently dominate the Dutch military. In terms of Moskos's theory, the most relevant question pertaining to this mood is, "Where does the institution come in?" The institution is the underlying idea and binding force that holds the military together and gives it its meaning and purpose.

We will have to look for this institution among the bits and pieces of the Dutch military, and I will present those bits and pieces in what seems to be a logical order. First, I will look for the institution in recent Dutch history and present-day Dutch society. Then I will look for it in conscript culture and among would-be volunteers, using the motives of the volunteers to present a "phenomenology of the institution." Finally, in the main section of this chapter, I will look for the institution in professional military culture, or in occupational culture for that matter, because at first glance the developments produced by Moskos's theory are present in the Dutch military. The military has an occupational surface where one looks in vain for the institution, but words, figures, and people aren't always what they seem to be.

I.

In recent Dutch history, the memory of May 10, 1940, is imperishable. On that day, more than a century without war was ended brutally by the invasion of the German army, as Dutch military and civilian defenses collapsed with the bombing of Rotterdam.

A historian once wrote that the Dutch army was seldom as popular as at the moment when it was taken prisoner of war. I do not think *popular* is a suitable term in this context, but it is clear and probably true for what the historian tried to suggest: after the surrender of May 15, a feeling of togetherness existed between the Dutch army and the Dutch people that had been absent until that day. It came into being as a result of the common defeat. When togetherness no longer mattered, society and the armed forces found each other.

Was the relationship bad before World War II? Yes, it was. It was bad before World War II and again afterward when the euphoria that was projected onto the liberating armies passed on. In the beginning, the euphoria had literally moved eastward toward the Dutch Indies. Dutch soldiers boarded troopships enthusiastically, determined to shake off Japanese tyranny on behalf of the beloved and no doubt grateful colony. When the troopships arrived, Japanese tyranny had collapsed already. Instead of liberating their colonial compatriots, the soldiers had to fight them for long years in the type of war in which the Netherlands broke its back, just as France and the United States would do in years to come. As J. A. van Doorn wrote, it was a miracle that a worn-out country, which did not even have time to recover from World War II, succeeded in keeping an army of 100,000 soldiers—volunteers, professionals and a majority of conscripts—in the field at the other end of the world. It was a motivated fighting force almost until the end.

When the soldiers came home in 1949 and 1950, however, the euphoria was long gone, and Holland had had more than its share of war. About 12 years later, this decolonization drama was repeated on a small scale in what nowadays is called West Irian. In 1975 Suriname became an independent country, and this time there was no war. In February 1980, Surinamese sergeants staged a coup d'état; they were Surinamese sergeants, to be sure, but they had received their education and experience in Dutch military culture. Moreover, the Dutch military attaché in Paramaribo may have done a little bit more than merely observe the coup. In addition, 3,000 volunteers fought with the UN forces in Korea, and Dutch soldiers—not necessarily volunteers—have carried out peacekeeping missions in Lebanon and the Sinai, just as they have acted on behalf of the United Nations in other troubled places around the world. At home, an elite unit of Marines made worldwide headlines in 1977 by liberating train passengers who had been taken hostage by South Moluccan terrorists.

These feats of arms are important, but nevertheless are footnotes in the post-World War II history of the Dutch armed forces, which have returned to being essentially a peacetime military. This military thinks about war, prepares for war, and simulates war, but does not wage war. This is a fundamental condition, although not exceptional. The pre-1940 pattern has been resumed, but in a different international setting.

NATO membership in particular seems to be significant for Dutch society and for the Dutch armed forces. Public opinion has always favored NATO membership, although probably without strong feelings. As a matter of common sense, the Netherlands cannot defend itself on its own. Military defense outside NATO does not have many supporters; the main political parties also agree strongly on this point.

During the last 25 years, public opinion polls have shown that at least 80 percent of the Dutch people believe a military defense is essential. Some individuals and groups hate the armed forces and can't be overlooked, but other people, some of them organized in associations, love their army, their navy, their air force. Public shows never fail to attract a mass of spectators, especially when warships or planes are involved. As long as the air force does not make a habit of flying above their hometown, a lot of people are impressed by these technical wonders. In the same vein, the further away the army tests its guns and tanks, preferably in Germany, the better the people like their army.

II.

The armed forces in the Netherlands have a peacetime strength of about 130,000—46,000 conscripts, 56,000 volunteers, and 27,000 civilians. These numbers are expected to remain stable in coming years. Nine out of 10 conscripts serve in the army. Thus, practically speaking, when we discuss draftees, we are talking about the army.

Slightly more than 10 years ago, national and international commentators often called the Dutch army an army of hippies, seldom in a flattering sense. Now that fashion calls for short hair again, how little Dutch conscripts have in common with hippies is clear. On the contrary, they like their compulsory military existence as orderly as they can have it. Their wage level is related to the legally guaranteed minimum for young people in civilian society. They prefer to work from 8 a.m. to 5 p.m., with proper breaks for coffee, tea, and lunch. Extra work (guard duty, exercises, and so on), means extra pay, especially after the initial months of basic training. If possible, they spend not only weekends but also Monday through Friday nights at home and go home as soon as possible. Immediately after 5 p.m., the conscripts jump into their secondhand cars and leave the barracks. This means that quite a few of them skip dinner, so their unions now want to eliminate the food deduction from wages. Why should one pay for food one does not eat? This point is being negotiated with the Ministry of Defense, as are many other conditions of employment.

Dutch conscripts are not hippies anymore; they are employees, watching out for their rights. Their employee mentality is always good for a laugh, and a comedian can score an easy success by referring to the attitudes of conscripts: "We ask the Russians please not to attack after 5 p.m. and during the weekend."

More serious commentators also worry about the conscript mentality. Some Dutch sociologists have questioned whether in the long run an army can do what it is supposed to do with soldiers who are the spoiled children of a welfare society. Moreover, the option of an all-volunteer force has been considered seriously by a government commission. Although the majority of the commission favors the idea, it concluded that recruiting sufficient numbers of volunteers would be too costly.

The core of the Dutch military—career NCOs and officers—have always favored a mixed force over an all-volunteer force, although they detest some of the more conspicuous expressions of conscript culture. Conscripts seem to be their favorite SOBs. A sergeant recalls:

> In the past as a company sergeant-major you literally could kick them around. They would even be grateful to you. Now it's become more difficult, you have to search for other solutions, but you learn. And it does have its good sides, because some noncommissioned officers really behaved like beasts. But then again, there was the fatherly type and that's the one I miss, now and again.

The captain, a company commander, comments:

> In my opinion 95 percent of the conscripts are of good will, as long as you give the right instructions, are clear to them, never leave them in doubt what they can expect.

During a NATO exercise, a general is impressed by his soldiers whom he calls "the Israelis of NATO":

> In all areas where brains, imagination and initiative matter, they beat the other nationalities by far.

For a general, of course, an international exercise had better be successful, and in public one will never hear him admit it was a failure. Nevertheless, the standard opinion is that especially during field exercises, Dutch conscripts do what they can and must do and do not neglect duties in any systematic sense. If they must, if the opportunity is there, conscripts will give what it takes, and that includes a capacity to endure hardship. Generally, being an employee at one time does not prevent a person from being a soldier at another time.

Whatever doubts exist about the quality of conscripts are related to the growing complexity of certain functions. The handling of a Leopard tank, for example, seems to be too difficult to master in 14 months. The driver is a short-term volunteer, and the gunner may be one as well. Studies are being conducted on the option of a differentiated draft period that would present the choice of serving longer and performing a more difficult, more interesting function. Whether this system will be realized is rather uncertain.

Whatever its modality, the draft will probably survive in the Netherlands, at least until the end of this century. It will definitely survive as far as conscripts themselves have a say; generation after generation, year after year, levy after levy, they go when called. Conscription works in our time, just as it did in the past. In fact, the biggest problem is how to get rid of the draftee surplus, because among those declared fit after physical examination, only half are needed. These numbers will change in the coming decade, but not to the point of shortages.

When given a chance, the draftees themselves will do what they can to cut the surplus. In their culture, not having to serve is a piece of luck, at least at first glance, and a lot of colorful stories are going around about how to avoid the draft. Getting the maximum points for instability is one of the favorite stories, but even if some actually try it and succeed, for most of them it remains a story about what one could do or would do or should do but in the end did not. Probably, most of the draftees declared unfit because of instability are truly unstable, or at least are expected to be a pain in the neck when in the army.

During the first months of service, unstable draftees show a high attrition rate. A study conducted some years ago among three infantry battalions showed that between 5 and 18 percent of the levy was declared unfit because of instability. In fact, *instability* is a vague term for the many different problems a person can have with the army, or problems the army can have with a particular individual. Mostly, of course, the problems are mutual.

Conscripts look for excuses for not serving, but if their luck fails them, they do serve and become effective employee-soldiers. Of course, they can also apply for conscientious objector status; roughly 2,000 take that road each year with success. Even so, the youth who has run out of luck and not troubled by his conscience will have to serve, and when you ask him why he is going, that is exactly the answer he will give: "Because I have to." He might add, "Otherwise they come and get you," or "If I don't go somebody else would have to do the job for me, and one way or another there have to be people serving in the army. Moreover, the majority of the Dutch people are in favor of conscription; it's a law. Besides, a country must defend itself. Otherwise we might just as well invite the Russians to take over. If I have to I'll fight, although I'm not sure nowadays a soldier with a rifle makes much sense."

With these kinds of statements, conscripts give meaning to the act of serving. In this informal logic of everyday life and language (I borrow the expression from Clifford Geertz), the underlying idea, the legitimacy of the institution is found. You must look carefully to find it, and in a moment it is gone again, but it definitely is there: the significance of the draft, the army, and military defense.

In addition to showing a lot of employee and occupational mentality, conscript culture has strong rhetorical overtones. These two phenomena succeed in keeping the army at a distance and hiding its institutional essence, but when necessary, this essence can be mobilized. Just as the army itself could be mobilized throughout Dutch history, just as its meaning now and again comes to the surface amid the hot issues of daily defense politics, and just as it is always in the background of public opinion, so in conscript culture the basic values can come out into the open, but only when they must. Normally conscripts have many other things on their minds.

III.

Would-be volunteers have other things on their minds as well. Listen to these two ambitious young men:

I've had my physical tests and I plan to join the professional military next year. Then I'll have this fellow coming to me and if he thinks I've got the right motivation, I think they will take me. I mean, if you succeed in becoming a pro you've really made it, you get paid well, you'll always be sure of having a job, you won't be fired, and in my opinion these are the things that count.

I feel attracted to the army, I like having to deal with other people, and besides, it's a rather steady job. You're assured of having an income. But of course, that's not the kind of motivation you can come up with, I'll have to think about that.

This last youth makes a charming confession, and probably he is right. He who wants to join the career military had better not talk only about having a steady job and a good salary. The level of military service makes a difference but generally volunteers are recommended to dress up their ambitions, although without being hypocritical; "steady job" and "good salary" are respectable motives. In a 1983 survey of sailors conducted by the navy, "having a steady job," appeared to be the first reason for joining. "Seeing the world" was reason number 2, and "because working in the navy seemed attractive to me," number 3. Also mentioned, but less frequently, were "sailing," "getting an education," "having a salary," and "working with technology." Some of these motives are seen in the following comment of a would-be sailor:

I've passed the physical tests, which are really something in the navy. Very strict, only a few boys come through. Once you've joined, you get an education and everything with the technical branch, computers, and so on. I guess you see a lot. A bit of adventure, a lot of travel, maybe to Curaçao.

An officer candidate puts his ambitions this way:

Probably I'll join the navy, the Royal Institute for the Navy, it is called. It takes three years, I think, and after that, well, I'll work as an officer. I've got a family, so it looks a real fine job to me.

I do not have any figures on the subject, but having a military family definitely directs some ambitions. The next volunteer is an example:

My whole family is in the military. My father is a veteran, my cousins serve, even my girl cousin. Some years ago my father was declared unfit because of the wounds he got in the East Indies. Personally I want to join the army because I really dig having a varied job, a lot of sports, and getting an interesting, focused education. And they pay you from the beginning of your education. You can be certain to have a nice job, right after you've finished your education. There's a lot of good comradeship, but still there's room for your own ego. And it's fun giving instructions to other people.

With the exception of the sad story about the father, this almost sounds like a recruiting ad. "A lot of sports," "a focused education," "comradeship but still room for your own ego": one would almost like to sign up immediately. In the meantime, he demonstrates no hypocrisy about a steady job and a good salary. Enthusiasm for joining can be so great and so deep that giving the reason becomes somewhat difficult:

It's strange, I really can't tell why I'm going to join the army. I mean, it's not a calling, not that I'm driven on behalf of country and people, not at all. I'm rather the egoistic type, I see it as my profession, where I can feel at home. I mean, I want to be happy and to me this looks like the job for being happy. I don't know. I've always wanted it, from the time I was four years old, I guess. Something I do like very much, that I must admit, I like things going in a disciplined way, smoothly, on schedule, and of course that's the army. Well, maybe the best way to explain it is what I usually say when people ask why. I always tell them, if the Russians come and get us, I'd join up with them, I don't care.

I always wondered whether he used this explanation when he actually applied to become an officer and how the committee reacted. If they listened carefully, they would know that in the context of everything he is saying, his comment about the Russian army is practically a declaration of love. He has a love affair with the army, any army, as long as it has its special army quality.

Evidently the rhetoric of my "country" and my "people" is not necessarily a part of this love affair. It may be a hidden part in the case of the next would-be officer:

In my opinion the army is spending a lot of time in the open air, being busy with sports, being some kind of a leader. So then you'll have to go to the Royal Military Academy. To me the army is a wonderful organization, everything always happens just at the right time. If they have to go someplace, you can bet they will be there. That's my style. You know, I'm not the patriotic type. People say you are defending your country, aren't you, and of course, it's all for the sake of your country, but that's not your motive for joining.

This statement suggests that the why of signing up can be separated from the why of the profession, at least the professional why that lies outside the everyday working world. In this way, the underlying idea of the institution takes its place again, very much like the place it has in history, in society, in conscript culture. Of course, this handful of quotations is not in any way representative of all the professions, occupations, and jobs the military has to offer at so many levels; neither are they representative of all the men—and the few women—who actually apply for them. However, they do suggest that the institution has a rather hidden status in the informal logic of everyday life: it definitely is there, it can be mobilized, but usually a mass of other motives is in the foreground. To give one more example:

I always have wanted to become a professional, an officer. you know, a tight group of people, working together, people with whom you have a good relationship. Discipline, of course, there has to be discipline, but not like it used to be in the past. Just a team, You know, which also cooperates with the other services, in order to serve the Netherlands. And to safeguard the peace, there's no need for getting aggressive against the other party. Taking care of the peace in your own country, but in other countries as well. The UNIFIL military is doing a fine job in Lebanon.

Here we have that second pillar of meaning, *peace*, which together with *country* can give service in the armed forces an aura of idealism. It emerges usually only after other motives have been expressed, but surely idealism is part of *applying* for and part of *doing* the job. In 1974, when a sample of the professional military was asked, "Do you think serving in the military requires a dose of idealism?" 90 percent answered yes. That percentage suggests that some social desirability is involved. We do not doubt their sincerity, but seeing precisely where and when idealism enters the picture is important.

IV.

By no means do military volunteers have a steady job in the Dutch armed forces. Roughly 15 percent of the personnel sign up for a limited period. They belong to all ranks up to the level of junior officers and are found in the three main services. They serve up to a maximum of six years, and they know that they'll have to leave after having completed this period. Usually they cannot stay on for a longer period, certainly not in the same place and the same function; the short-timers know this from the beginning. At least, they're supposed to know it, but all of them may not realize this fact with all its consequences. The government and the state bureaucracy are discussing whether these short-timers should have first choice in getting jobs as public servants. In the army, a category of low-level technicians (drivers and mechanics) have the opportunity to acquire a civilian diploma. Not everybody is succesful in acquiring it, nor does it guarantee a job, which depends on the labor market. Nevertheless, the position of these technicians differs from that of the short-timers who must rely on their own initiative to plan their postmilitary career.

The government has also questioned whether volunteers' cash bonuses, which they receive at the end of their term, should be maintained. At the present it has been reduced, perhaps as a first step in doing away with it. The policy of the Ministry of Defense seems to be, "Why pay them a bonus if we can get them anyway? If necessary, we can always give bonuses for categories of personnel that otherwise are hard to recruit." This point is especially relevant with regard to high-level technicians, for whom the armed forces must compete with big industry. Money may well be the only way of attracting and keeping them.

The scarcity of high-level personnel, however, is quite different from the problems of low-level short-timers. In a social and psychological profile, their motivation for signing up is portrayed rather negatively. Youth unemployment and the escape from personal circumstances supposedly are the dominating motives of these youngsters for choosing the armed forces, which are a kind of refuge. To complete the gloomy picture of these short-timers, they are irresponsible with their bonus, they sit out their time, and afterwards they are as unemployable as before. A few quotations from volunteers suggest that the way these volunteers are viewed is not just an idiosyncrasy in Defense Department circles, but a commonsense story as well:

Most of the volunteers I know had no other option. At their educational level and there are so many of them—I mean literally thousands—there's no way of getting a job.

Today a friend of mine went to the army. He volunteered. He joined the tanks to become a tank driver. He dropped out of school without a diploma, had a lot of jobs, got fired, and so on, so that's why he signed. He likes it, he's really happy.

These stories emphasize that official concerns and worries have a basis in reality. Nevertheless, at least in Dutch circumstances, this short-timers' profile sometimes seems to have been inspired a little too much by middle-class values about how one should plan and organize one's life and career.

The primary concern of the armed forces is not whether these short-timers will make a success of their lives, but whether they are good at doing what they're supposed to do while they are in the army. Generalizing on this point is hard, not only because of the heterogeneity of the category, but also because not much research on this subject is being conducted or published. The small amount of available data suggests that low-level short-timers are bad soldiers but good technicians—bad at traditional soldierly duties and good at driving and repairing tanks and trucks. Exit interviews conducted among these short-timers show the same picture: they were more positive about the technical side than about the military side of their army service.

Some commentators have stated that short-timers perform worse among a majority of conscripts, who notoriously dislike all disciplinary behavior. When not in direct contact with conscripts, they supposedly perform better. This idea sounds convincing and also suggests a problem common to short-timers: dispersed as they are in small numbers among a majority of conscripts, they lack their own subculture. This situation is said to cause the short-timers to suffer some attrition, that is, to leave the military before their term of service ends.

Attrition because of a lack of subculture also applies to women in the armed forces. In the Dutch military, women are a very small minority, and their numbers cannot form much of a subculture. Less than 3 percent of all the voluntary personnel are women, 3.5 percent in the navy, 2.5 percent in the air force, 2.4 percent in the army. In addition, or partly because of this small proportion, their attrition rate is much higher than that for men. In 1982, 11 percent of the women left the army before completing their turn; the overall attrition rate for men was 1 percent. Women leave sooner and have more difficulty getting in; relatively fewer female applicants are accepted, and fewer females apply in the first place. The Dutch armed forces are far from being a mixed organization, a fact that cannot be hidden by symbolic successes like the first female pilot or the first mixed warship. The Defense Department has been discussing offering women extra-short-term contracts, possibly two years, with a possibility of extension, as a kind of introduction period, such as men experience in serving as draftees. Actually, two years is not much longer than the 14-month draft period, but the women would be paid just like the other volunteers.

For years, conscript unions have pleaded for conscripts to occupy the same pay level as volunteers. Their motto is "same work, same pay." Thus, when the plans for extra-short-term contracts for women were proposed, the conscript unions were predictably angry. Women seemed to have what the conscripts had always wanted but could not attain. In fact, the conscripts should be happy if they succeed in saving the minimum wage for young people as a standard; rumors have the Ministry seeking a less costly standard. Whether all this will become reality is still unknown, but taken together, we see how conditions of employment are weighed and discussed in the Dutch military and indeed, the mood of the military. Among a lot of other things, money counts, and not only among conscripts, short-timers, and women.

"If I had wanted to make a lot of money," the general tell us, "I would have gone to big industry. In fact, I've had some offers, really for a lot of money. In my present function I am the head of 10,000 civilians and 70,000 soldiers. I have a budget of 5.4 billion Dutch guilders. With such responsibilities I could earn far more than the 150,000 guilders I get at this moment. But I am satisfied. Money is not what keeps me going. To me money is no stimulation."

Thus the general, the commander in chief of the land forces, did not give in to the seductions of big industry. Others did, and do. The *Officers' Association Monthly* tells us: "In the last couple of years tens of those recommended for top-level functions—accountants, legal specialists, automation experts—have quit." The editors are deeply concerned about what they see as a "brain drain" that causes serious problems for the armed forces. They do not know exactly why these people have left but have their suspicions: too few possibilities for individual career plans. The best people supposedly don't have the opportunity to move up faster into the most challenging jobs and receive adequate pay. Now, as a rule, say the *Officers' Monthly*: "Everyone is being paid according to rank and years, no matter which function he is fulfilling. Being better and working harder does not pay off."

The Defense Ministry denies any serious problems with turnover rates of high-level personnel, but the official *Defense Journal* states the general motivation for quitting as "higher salaries and better secondary conditions of employment elsewhere." Thus money is involved; the officers and the Defense Department agree at least in that respect.

Money definitely counts in rumors about an exodus of air force and navy pilots to civil aviation. Pilots earn highest extra pay that the armed forces provide, but this is nothing compared to what civil aviation pays, and evidently the civilian salary is seductive. Some turnover, more or less regulated, has always existed, but now the number of pilots who are supposedly thinking about quitting is excessive—more than civil aviation could absorb. As usual, the ministry denies any serious problems.

Clearly, turnover rates can function as a kind of pressure. Unions and associations are negotiating their conditions of employment with the ministry by saying, "If you don't pay us better, we'll quit." In fact, a recent official study showed that the military is relatively underpaid, considering all the extra work military people have to perform. The study did not even consider big industry, but compared civilian and military employees of the armed forces.

A 1984 survey among career sergeants and corporals who quit the navy after fulfilling their first term showed general satisfaction with pay levels. The navy pays well, 72 percent said, but a majority judged compensation for extra work and other inconveniences as unsatisfactory, a finding that agrees with the above mentioned study.

Several surveys have shown, however, that the most important motive for quitting is a combination of complaints that center on too many assignments at sea compared with shore assignments and the effect sea duty has on private and family life. This is the dominant complaint of noncoms, corporals, and sailors who left the navy.

The ratio of sea to shore assignments, and the consequences, are not very important to officers who leave the navy. Officers generally sail less often, and they quit later in their careers and their lives. Family problems caused by absent fathers usually arise at an earlier stage, according to the sociologists. Most officers leave the navy because of the "kind of work they are doing" and "the availability of a nice job elsewhere." In sum, officers leave because/and when they can find a job they like even better than the navy. "Kind of work" stands for factors like "responsibility," "challenge," or "career prospects." Salary is not mentioned frequently as a motive.

Attrition and turnover in the military have a somewhat peculiar flavor. They are generally not accepted as normal. On the contrary, they imply a problem, either on the side of the armed forces or on the side of those who quit. Judgments such as "Choosing a military career is no longer as it was in the past, a matter-of-fact choice for life are not unusual." One of the conclusions of the navy surveys is that serving in the navy is seen more and more as a youth occupation and that "the younger people have less of a bond with the Navy, which they do not see as a way of life, but as just another job." One should be careful in mythologizing the past, in which "real bonds" existed between military personnel and their beloved organization. Moreover, why should hanging around for all of one's working life be the best proof of worthy motivation? People can perform well in the armed forces, even if they do not stay throughout their working lives.

In short, some degree of turnover should probably be seen as normal, although it can cause problems in the organization when too many of the same kind of people leave at the same time. These problems could be tackled by horizontal recruiting. This is already being done by taking officers and noncommissioned officers in their thirties who have some military experience as conscripts or as short-timers. They receive a shortened training period and then are placed in middle-level functions with the appropriate rank.

This policy is pursued when personnel shortages occur because of earlier budget cuts, a skewed age structure, and turnover. Naturally, personnel who had to start at the bottom are watching closely, by way of their unions and associations, to see whether horizontal recruiting damages their careers. Could one imagine horizontal recruiting at top levels? Most of the Dutch military do not see this as a possibility. One infantry lieutenant colonel, who writes emphatically about "military management" and is prepared to use civilian expertise whenever possible, rejects horizontal recruiting at the top. "We cannot imagine a top-level manager from Philipps who becomes a commander in the military." Maybe in the economic or financial branch of the armed forces such a transfer is possible, admits the lieutenant colonel, but definitely not in real military functions. Here, the military needs those who come from the inside.

V.

Finally, we'll take a look at the solid core of the armed forces, those who enter during their late teens or early twenties and stay until they retire in their midfifties. (Then they can often have a second career in a civilian environment. The navy retires its personnel at 50, the air force and the army at 55. Plans have been made for an overall retirement age of 58.) The idealism and attachment of career army, navy, or air force personnel is not in doubt. Survey data show such idealism and attachment. However, where and when do idealism and attachment come in? What exactly do they stand for? Certainly they do not stand for a complacent attitude that gives the organization freedom to do what it likes, regardless of hardships and inconveniences.

As we discussed earlier, the Dutch soldier wants the extra compensations that are part of the job, but not all the inconveniences can be made good by money. This conclusion comes from a 1984 survey in which a sample of the professional military weighed a long list of inconveniences. Infringements on private life and violation of privacy were disliked most. Lodging without privacy, living in a military neighborhood, being at home only during the weekend, having to move, and being sent abroad without family were all listed as major inconveniences. However, the Dutch professional soldier did not care about not being allowed to strike or about being subjected to military law.

"Having to risk your own life" isn't seen as much of an inconvenience. On a list of 41 items, it ranks 22d. "Having to use your weapons while on duty" ranks even lower at 36th. These results suggest how much of a peacetime force the Dutch military is. In the everyday world of the Dutch soldier, the idea of actually having to fight is far out of place. In comparing inconveniences, items such as "lodging without privacy" weigh much more heavily.

The professional soldiers would like an opportunity to spend more time enjoying life outside the armed forces. At the moment, the organization does not allow them to use all their free days, because even if extra work was done after hours, organizational goals for exercising personnel and equipment could not be met. Thus, some of the free days are never used. This suggests that when given no alternative, soldiers accept the inconveniences more or less inherent in their profession: they work when they would like to have the day off, spend their weeks somewhere else when they would like to be at home, and move when they would rather stay. If necessary, they even move to a military neighborhood. Most likely that neighborhood is in West Germany, where a Dutch army brigade and a couple of air force missile sites are permanently encamped. Whether their stays are successful seems to depend on the adaptation of their families to the new environment. A recent study of military families in West Germany recommends considering family factors before sending soldiers on German tours. From the attitudes and motivation of the spouses, the study concludes, some predictions can be made as to the probable success of their stay.

One sergeant and his wife have something to say about the relationship between private lives and working in the armed forces.

> They put me in another function because I had been doing staff work long enough. But I want to go back to the staff, because I want a job from 8 till 5. All those exercises, night shifts, guarding duties, all those extras, I simply don't want them. My wife says, "It's ridiculous; you're not paid to do them." I could use that time a lot better, by running my little insurance business. Yes, a lot of military people do something like that. And then I haven't any time left for my hobbies; my whole social life is a mess.

What the sergeant says has rhetorical overtones because while he speaks he is with his unit in the midst of a field exercise. Nevertheless, one hears this kind of story a little too often to dismiss it as pure rhetoric. Colorful stories about transfer policies and how to avoid them echo around the armed forces. One cannot help being reminded of the conscripts' stories about how to avoid the draft.

More serious is the tone of the personnel functionary who analyzes the problems inherent in transfer policies. He confesses his dilemma of being in doubt about the wisdom of that policy and nevertheless having to work with it. He must advise about the transfer of people who do not want to be transferred and who even might simulate social and medical problems to avoid transfer. Nowadays, even the prospect of a career does not always seem to make up for the consequences of frequent relocations; some people choose not to have a career. Again we should be careful about mythologizing the past because these stories have been heard before, just like the turnover stories cited above. Nevertheless, talk of not having a career and of giving priority to private life has wide currency. Implicitly, such objections confirm that having a career and being moved around still set the tone and can usually be expected. Because of that, anyone who does not want to have a career draws attention. The sergeant major who refuses promotion to officer because he is happy in what he is doing and fears the stress for him and his family when he moves up will certainly be talked about. So will the young captain who puts aside his recommendation for top-level positions because he loathes the removal policies to which he will be subjected as a result of this recommendation.

Then, too, the captain's wife has her own career, which would be jeopardized if they had to move. Moreover, top-level functions do not pay that much more, not enough to make it all worthwhile. Of course, when a lot of captains refuse a career, the organization has a problem. I would imagine, however, that many more soldiers are dissatisfied because they do not get the chance to move up, even if moving up in the armed forces presupposes moving self and family inside or outside the Netherlands.

This section covers many important issues, each of which deserves more attention than I could give it. Taken together, however, they contribute to the picture of the Dutch military. Is it a "real military," one might ask, in view of all this emphasis on privacy, reluctance to be moved, clinging to functions and specialties? It is a very real military, I would contend, if only we could agree on what a real military is. As the next section will show, this consensus may be difficult because even the insiders themselves do not agree on what a real military is.

VI.

Of every three professional soldiers in the army, one is an officer and two are noncommissioned officers. Of the two NCOs, one is an instructor and the other in an administrator, a manager, a technician, or another kind of specialist. Both NCOs receive the same basic training, which is shaped much more by the image of the instructor than by the image of any specialist. *The* NCO, in fact, does not exist, and this remark has become a cliché in its own right.

An expert in the structure of large organizations, who teaches at the Royal Military Academy, searches in vain for some civilian counterpart to the NCO corps. He wonders whether the corps, as something separate, could not be eliminated. At the same time, the officer corps would be deleted because the NCO corps is the artificial boundary between two kinds of personnel. Certainly upward mobility for sergeants would become easier, and some traditional frustrations and tensions would vanish. So now and again the sergeant himself, or the NCO association, likes to flirt with the idea. However, doing away with the NCO corps would remove a sense of community as well. A professional image that binds NCOs, that creates a subculture and a set of identification symbols, would vanish. The NCO rhetoric would also vanish; this rhetoric gives the sergeant an opportunity to pose as the real military man who has his heart in his job and is the stable element between ever-changing conscripts and ambitious young officers who are looking upward and forward. The sergeant, the army's backbone, likes to say, "They have the careers, we do the job."

In the meantime, the captain has some serious problems in behaving like a leader, and everybody, including the general, agrees on the cause. The company commander has far too much paperwork, which keeps him at his desk and away from where he is supposed to be, among his people giving instructions and inspecting equipment. In fact, he does manage to instruct and to inspect, but only by working long, busy days that go something like this, the insider tells us:

> A company commander has responsibility for about 150 men, equipment, weaponry, training program field exercises, preparing war. From 7:30 till 17:30 he is busy: about once a week he has an evening or night exercise; all in all during one year he spends about four months away from home; he constantly has to check equipment and equipment maintenance. He lives under the stress of being supervised and judged himself, but he also tries to help his commander receive a favorable judgment at official inspections. Moreover, he has to have an open eye for personal problems his men might have. And let's not forget, when the weather is bad there is no flying, but there are definitely field exercises.

This last remark gives us the context of this quotation: a discussion in the *Officers' Association Monthly* about the complaints and wishes of air force pilots, which angered some of their army colleagues. The officer I quoted informs the pilots about the burdens of being a company commander who does not get an extra 17,000 guilders each year. On top of that, the army officers grumble, pilots are not even willing to have an ordinary military career or to do their share of staff work. Well, the armed forces definitely do not need people who can only manipulate a joy stick or who have an elitist mentality. Those pilots had better realize that they would not be in the air for a minute were it not for the ground personnel. Pilots respond by pointing out the extreme physical and mental requirements they have to meet. Very few people pass the tests. Pilots also have to cope with "brief moments of sheer terror" when they are in the air. Then they mention civil aviation, where air force pilots can make a lot more money. Company commanders, the pilots contend, are not much in demand in civilian circles. This last remark probably proves to the army officers that pilots do not belong to the real military.

Also in demand in civilian circles is the administrator accountant who is not much of a real soldier either, according to colleagues.

> The officer of military administration is seen less and less as "one of us," a feeling stimulated by the power pencils seem to have in peacetime armies instead of guns. In the eyes of the colleagues this is proved by the fact that the branch of Military Administration counts five general and twelve colonels. In proportion to other branches and services this is seen as excessive, so they call us the "pink Mafia." [Pink is the color of the paper the administrator produces.]

Meanwhile, the administrator has a nice idea about how to help the company commander get rid of paperwork: a new personal computer with a hard disk, a floppy disk, and a printer. Does this jargon prove the administrator might as well be a civilian? No, that would be jumping to conclusions, because, as the administrator says:

> If the work of a gunner in a Leopard and a pilot in an F-16 is made easier by automation, why

shouldn't the same thing be done for the work of a company commander, and his sergeant major in administration?

Indeed, if the use of computer jargon were a criterion, hardly any real soldiers would be left. Many of them would be exposed as civilian like experts and specialists. Anyway, the Dutch armed forces already are employing an impressive number of civilians. In the eyes of the real military, they do not have much status. This does not necessarily preclude good teamwork and personal appreciation, but generally the soldier comes first. The career pattern of the armed forces gives priority to military personnel. At the top of The Hague bureaucracy, civilians have nice careers, but far away in the country, at a place where tanks are being repaired, the sergeant major, not a civilian, supervises the military and civilian personnel. For a long time, civilian employees felt that their conditions of employment were worse in respect to other career patterns but, as I said earlier, an official study group seems to have proved that military personnel are underpaid. This finding must have come as something of a surprise to the civilians; one wonders whether they really were convinced.

The eagerness with which professional soldiers claim to be the real military, or the most vital element, to the exclusion of other branches and specialties, makes one think about that famous cornerstone of professionalism—corps loyalty. Which corps? There are so many. And what about that other cornerstone, expertise? Which expertise? There are tens, maybe hundreds.

Professionalism has yet another cornerstone: being in the service of the country, the state, society, the people. Indeed, this factor seems to have some binding force for the Dutch military. Sometimes, however, the binding seems to be done not by serving, but by resisting all of those nosy outsiders.

VII.

"Not invented here," in English, is the slogan Dutch soldiers use whenever they are confronted with measures that intervene in their management of the armed forces. That kind of intervention also comes from the inside, from top to bottom and from staff to troops, but "not invented here" is used normally in relation to real outsiders—civilians, especially politicians. National politics are the source of all those things that the military does not want. As one lieutenant colonel said, "'Not invented here' usually means 'not wanted here,'" which applies especially to civilianization and all its perceived excesses, from untidy-looking conscripts to intricate participation procedures that may or may not produce democracy but certainly produce extra paperwork. For example, consider how disciplinary measures must be taken. First the charge must be written out; then the accused soldier and his self-appointed confidant must be heard; and finally the sentence must be written out. This is extra work, which the commander certainly did not invent himself. Moreover, the spectrum of penalties at his disposal has diminished. The commander no longer has the authority to sentence personnel to close arrest in the barracks, let alone to a military jail. Intead he has become a specialist in fines.

The latest invention of "politics" is the integration of women into the armed forces. This is not much of a numerical success, as I said earlier, but it attracts a lot of publicity. The first mixed warship was an item of interest to journalists, sociologists, feminists, and government advisory boards. The navy itself was not amused, certainly far less than when it heard the future Dutch king, who had reached draft age, was planning to serve as a reserve naval officer.

Princes aboard, yes, but women aboard, no. A survey among sailors showed that 75 percent were opposed, and this opposition goes all the way to the top. The higher one goes, the more solemn the arguments sound, and if someone is in the mood, one can hear him argue that the balance of power is at stake.

Seldom, does someone say, "What you invented in The Hague does not damage our fighting capacity, but surely it threatens our way of life, our all-male subculture to which we are attached and that we value as something in its own right." The arguments always sound much more businesslike; they always point out the supposed damage to the performing of tasks, which in the end is the capacity to fight. This is usually the line of argument taken by the military when confronted with what is neither invented nor wanted here. Attachment to a way of doing things and dealing with people is presented as a professional judgment on fighting power or at least as some kind of related measure. Now and again, the military might be right, but as a rule, one had better be suspicious. In the meantime, mixed ships are still sailing. On all levels, the navy is trying to live with the integration of women, as the army and the air force are doing.

Anybody can pose as an expert with regard to human relations, but who can judge better than the military insider the best engine for a frigate? Is the professional soldier who lobbies and fights for the best engine to blame when the politicians decide on another one for the sake of economic compensation orders? For years the armed forces had to drive, sail, fly, and shoot with worn-out equipment. Now that everything is finally being brought up-to-date and made ultramodern, budget cuts on energy, munitions, and repairs create new frustration in the military. Just as Dutch soldiers like their human relationships the way they were in some respects, they like their equipment the way it is going to be. Professional soldiers have been said to want their wars short, clean, and decisive. Today, I would like to add, they want their weapons ultramodern regardless of cost.

The frustrations of the military, as caused by what the soldiers see as unwanted political interventions, are not be underestimated. At regular intervals, the soldiers' associations warn that loyalty, motivation, and satisfaction are being threatened. Fighting power as well, they never forget to add—especially fighting power. A famous survey held by the Officers' Association some years ago, which detailed a great number of complaints and worries, was called "Endangered Fighting Power." In this way only very suspicious persons would have the idea that the associations were fighting only for the interests of their members.

In fact the associations do practice double-talk, and they know it themselves. Most of them state explicitly that they have a twofold aim: negotiating conditions of employment and caring about the quality of the armed forces. The officers are somewhat more outspoken than the NCOs, and when they discuss the quality of the armed forces, one can't always be sure whether they are not in fact lobbying for better conditions of employment.

Nevertheless, Dutch military associations perform an important function. A lively culture of associations, magazines, polls, and protest meetings certainly is not new in the history of the Dutch military, either for officers or for NCOs. This culture, however, seems to have gained new relevance. Now that the military, and the military profession, are in bits and pieces, the associations play an integrating role. In some respects, they are a binding force, and as a trade union that deals with politicians, they probably would be hard to replace.

In addition to problems with politicians, the military also has trouble with those whom the politicians represent, the Dutch people, who supposedly are not very fond of their soldiers. The survey I quoted several times above asked: "How often do people criticize you in relation to your being a member of the armed forces?" Never, answer 58 percent of the navy, 45 percent of the army, and 50 percent of the air force. Now and again, say 37 percent, 44 percent, and 42 percent, respectively; often, say 5 percent, 11 percent, and 8 percent. "Have you ever been abused or threatened?" is another question. Never, respond 82 percent of the navy, 73 percent of the army, 82 percent of the air force; now and again, 13 percent, 17 percent, and 14 percent, respectively; more than three times, 5 percent, 10 percent, and 4 percent.

Some fear that the issue of nuclear arms could be projected onto the armed forces and its members and drive antimilitary feelings beyond their normal level. Nothing of the kind has happened, however. Public opinion of the armed forces has not been changed by public opinion of nuclear weapons. But the armed forces have not remained totally untouched by the debate; surely, the military has its own debate, at times as emotional as the civilian debates.

Although nuclear weapons have made themselves felt, it is their lack of impact on the military that is striking. Under this supposedly new semantic layer that has spread itself over society and armed forces alike, business is going on as usual.

In discussions and studies of the Dutch military, the idea of the vanishing soldier has received much attention. Nevertheless, after seeing what they do, hearing what they say, and reading what they write, one must conclude that the *persisting soldier* better describes the essence of the Dutch military as a finite province of meaning. Have we found what we are looking for—the armed forces as an institution?

In the last few paragraphs, I have treated the Dutch career military maliciously for their internal rivalry over the phrase *real military* and for fighting against unwanted outsiders who invent and intervene. Yet whatever else is at stake in these fights—attachment to the way human relationships were, eagerness for the weaponry that is going to be—a professional mentality is at the heart of it all. It represents the straightforward mentality of people who are experts at something and who want to have room to use their expertise. The Dutch military contains a lot of people who are experts at a lot of things, from the simultaneous handling of conscripts and company commanders to the engineering of high-technology frigates and F-16s. Nevertheless, at the level of the military as a collection of profession and occupations, professionalism sets the tone, leads the way, and binds the multitude of jobs, just as occupationalism appears to dominate at the level of conditions of employment in the military. Developments in conditions of employment seem to be unequivocal in the Dutch military, and in regard to the soldiers and their unions and associations, the end of this occupational "hard cash" trend has not yet been reached.

On the level of the military as a community, a way of life, and a subculture, developments are more equivocal. On the one hand, a great deal of emphasis is placed on privacy and on the separation of work and private life, not only physically but mentally as well. On the other hand, while at work, the Dutch military still shows some subcultural characteristics: having its own legal system, being practically all male, having its own formal hierarchy of ranks and groups of ranks, and having informal and idiosyncratic ways of doing things and dealing with people, including politicians and civilians.

"Very occupational on the surface, rather community-like inside, truly professional at heart" would be a summarizing judgment of the Dutch military, to which must be added, "and purely institutional deep down."

The institution as an underlying idea that gives the armed forces meaning, sense, and legitimacy is something to be mobilized. Ironically, the incidence and the relevance of this mobilization do not seem to increase, as one might expect, among the career military. Professionals have too much at stake in terms of their everyday lives and work. Professionals are too busy running the military and taking care of it. They can and do mobilize the institutional spirit, but to do so may sound pompous, or out of date, or all too fashionable, or like an exercise in social desirability. They are also suspected of being far more interested in something else—salaries and careers. They do make a living in the armed forces, don't they? They are making careers, aren't they? If they can make even more money somewhere else, they will quit, won't they? This is the ironic condition of the professional soldier, who is an expert at running the military and has not much of a chance to elaborate on its meaning. Then again, maybe that is how things should be, because we are forced to look elsewhere for institutional meaning—not in the armed forces, but in society. This also explains why the trends predicted by Moskos's theory apply in general to the

Dutch military, with some local peculiarities, but do not cause an erosion of the underlying values or of legitimacy. The institutional meaning of the Dutch military is hiding in Dutch society, but when it is mobilized, it appears to be alive and doing well.

Finally, this point can be illustrated by examining the word *mobilization* in a literal sense. If we take the army as an example, 30 percent of its personnel are on the alert and 70 percent belong to the reserve force, to be mobilized in times of emergency. These 70 percent are all civilians, teachers, farmers, factory workers, bus drivers. They are soldiers only at intervals, and in a peacetime army this means only when exercises are at hand. One can be a civilian for most of his life, for most of the year, and become a soldier when the moment is at hand. In a way, *mobilization* also applies to conscripts, short-timers, and professionals, the many men and the few women who are on the alert. The fact that they behave like civilians for part of the day should not deceive us; if they must behave like soldiers, they will fight, not only from 8 AM to 5 PM, but also after hours. The meaning of the military leaves them no other choice. In this sense, the institution lives on and on and on.

XV

Greece

DIMITRIOS SMOKOVITIS

In broad terms, the Greek armed forces have been very institutional, almost archaically so, but in the 1980s developments occurred that began to break the monolithic institutional mold. The first signs of incipient occupationalism, not all negative, could be discerned.

The context of the armed forces in modern Greece has been shaped by three paramount factors: interventions in the civil order, frequent involvement in war, and confronting military threats from neighboring countries.[1] These factors have profound consequences for the internal social organization of the Greek military. Since the turn of the century, a half-dozen direct or indirect military interventions in the civil order have taken place. The most recent was during the 1967–74 period, the era of the "colonel's junta." Also in this century Greece has been involved in numerous wars: the two Balkan Wars of 1912–14, World War I, World War II, and the Civil War of 1946–49. The Greek military has also engaged hostile combat forces in Korea (1951–52) and in Cyprus against Turkey (1974). Greece also occupies a vulnerable strategic position.[2] Potential enemies exist among the Communist countries on its northern borders. More daunting is a potentially hostile Turkey on the east. The past decade has seen a general shift in deployment from the northern to the eastern border. Because both Greece and Turkey are NATO members, Greek relations with NATO have become strained. On top of this, Greece feels psychologically "alone" by not having any permanently friendly neighbors or possessing cultural and linguistic similarities with any other European country.

Several historical conditions operate to foster a strong institutional form of military organization in Greece. The Western European countries—in turn, France, Germany, and Great Britain—from which Greece took its own models of military organization were themselves very institutional. Also, because of its lack of natural resources and indigenous wealth and its semi-industrialized nature, Greece has not been able to afford a highly technical military.

249

Perhaps the single most important factor has been the institutionalization of conscription since the nineteenth century. Although some small-scale opposition to conscription developed in the 1980s, the military draft still enjoys overwhelming support in contemporary Greece. The total regular strength of the Greek armed forces is about 200,000, of whom 150,000 are conscripts serving for a period of 22 months in the army (24 months in the air force, 26 months in the navy).

THE INSTITUTIONAL BASELINE

The basic time reference for the discussion of institutional and occupational (I/O) trends in the Greek armed forces is the period of the 1950s through the 1970s. Later developments are covered in the next section. The Greek military was and is probably the most institutional organization of any in the Western world. Its basic legitimacy derived from being defenders of such values as religion, country, and family. The command hierarchy saw itself as a protector of the national honor. Nationalism of an extreme sort was the guiding ideology. Esteem from the general society was based on the notion that all soldiers were performing a national service, and societal prestige was strongly correlated with the rank one occupied in the armed forces. Civilians were also somewhat alert to the differing status of various commissioning sources, with graduates from the Greek military academy seen as the most prestigious. This is not to deny the coexistence of negative stereotypes of military service that emanated from the political left. In these and allied intellectual circles, military officers were viewed as undereducated, fascistically inclined, and blind supporters of American interests.

The compensation system of the traditional Greek military was also very institutional. Enlisted recruits received in-kind subsistence and compensation amounting to a few dollars a month. In point of fact, the remuneration given draftees was not termed a *salary* but *pocket money*. All recruits lived in barracks and no private transportation (including bicycles) was allowed. Salaries between draftees and career noncommissioned officers were extremely decompressed, with senior sergeants earning about a hundred times the wages of a recruit. Because of the absence of any disposable income, draftees perforce spent almost all of their nonduty time with fellow draftees. The only exception to this pattern occurred among those draftees who came from more affluent families (that is, families able to send spending money to their soldier sons) and often associated with civilian youth of the same class background in the local area of their assignment.

Junior officers received about $100 monthly, and generals earned about five times that level. Small allotments were given for family members. Specialty pay did not exist, even for physicians. With no family quarters on base, officers had homes on the civilian economy. A cooperative and partly subsidized program allowed career officers to purchase homes. Certain sections in some towns, however, might be exclusively inhabited by military members. The primary career benefit was seen as a free and first-class (by Greek standards) system of medical care for self and family.

As in other institutional militaries, the orientation of military members was vertical within the military rather than horizontal to external civilian groups. A soldier was considered always on duty and was required to wear uniform even when off duty. Through the 1950s, senior officers even passed approval on the marriages of junior officers. Primary role identification was to the military service rather than to other social groups in the civilian society. For the career force, few friendship ties (except with relatives) were made with civilians. Social isolation was fairly complete and reinforced the anticommunist proclivities of the career force, most of whom came from the lower-middle class.[3]

The wife's status in the traditional Greek military was directly dependent upon her husband's rank. Spouses had no real responsibilities in the military community other than being sociable at officer parties. An officer's wife was judged by the criteria of being a good wife and a good mother. In reality, the military wife was a kind of appendage to the officer's military role, not someone with an autonomous role in the military community, much less a person with independent standing.

The military legal system had almost complete purview over the military member. All crimes, whether connected with the military or not, were subject to court-martial. Even off-duty traffic offenses would be adjudicated by a military court. The only exception to the total institutional control over the soldier's legal life was that certain civil cases involving a military member and a civilian (for example, an inheritance dispute) would be decided by civilian courts.

In brief, the Greek military in the modern era was almost a pure type of institution. Recently, however, significant trends have been moving the Greek military away from an institutional format.

RECENT TRENDS TOWARD OCCUPATIONALISM

The post-junta period since 1974 has witnessed major changes within the social organization of the Greek armed forces as well as in civil-military relations. Although the Greek military remains strongly institutional, developments indicative of occupational trends can be listed. These trends occurred within a political climate—first under Constantine Karamanlis (1974–81) and then, more so, under Andreas Papandreou (since 1981)—that sought to democratize the armed forces to preclude another coup, to reprofessionalize the military in order to make it more effective in the face of the Turkish threat, and to improve the living conditions of service members, especially junior ranks.

Compensation of draftees was modestly increased, but budgetary constraints prevented any real move toward anything approaching a genuine salary for the lower ranks. The buying power of a draftee's monthly remuneration was equivalent to about US $10. One minor change in conscript life-style was to allow draftees to wear civilian clothes when home on leave with their families. Certain radical groups in the civilian society agitate for a military draftees' union, but there has been no significant support for the concept. Although the end of the draft is not in sight in Greece, draft terms may be shortened.

The major compensation change in the post-junta era has occurred at the junior officer level. Salaries for junior officers approximately doubled in constant buying power since the mid-1970s. This development was caused in large part because of the declining number of qualified officer applicants. Indeed, without the high unemployment rate among middle-class youth, the officer shortage would be even more pronounced than it has already become. Since the junta days, a career as a military officer has suffered some tarnish and become a less desirable pursuit for young men, especially for those coming from the urban middle class. Compared to times past, a military officer apparently is not as desirable a marriage partner for women of that same class. The prestige of a military officer has come to be evaluated more in terms of compensation received, rather than on some abstract identification with an institution of national honor.

Another trend is that with more newly commissioned officers coming out of civilian universities (rather than the academics), the propensity to develop nonmilitary identities increases correspondingly. At the same time, wives of officers are increasingly likely to have outside employment, frequently professional or semiprofessional positions (something virtually unheard of a decade or two earlier). New nonmilitary patterns of friendship develop when officers interact with work colleages of their wives.

In the early 1980s, for the first time ever, a small number of women were allowed to join the Greek military. The participation of women in the armed forces in terms of new sex roles was never adequately tested. A controversy quickly developed over the fact that the women soldiers were given regular salaries (albeit modest), which contrasted to the pocket money received by male draftees. The resultant invidious comparisons on the part of the males, accompanied with general reluctance to employ women in any kind of military role, led to a governmental decision to phase out the program.

Perhaps the most significant trend toward occupationalism in the Greek military was the 1983 decision to recruit future technicians on a competitive basis. The plan was to recruit qualified secondary school graduates who would then receive technical training after entering the military. Such volunteer soldiers were recruited on five-year contracts. The plan was eventually to recruit 5,000 of the the technician soldiers. This was the first time in Greek history that enlisted soldiers were to be paid a wage commensurate with the private market. Recruiting the technician soldiers has, in fact, not been difficult. Data on the social attitudes of these volunteers is incomplete at the time of this writing, but some preliminary analyses indicate that such soldiers are much more occupational in their outlook than that heretofore characteristic of members of the Greek military.[4]

CONCLUSION

As long as the Greek nation faces threats from nearby countries, the military will maintain strong institutional features. Yet, the trend toward occupationalism within the armed forces is undeniable, especially in the officer corps and among the new volunteer technicians. A dose of occupationalism may be beneficial to Greece as an antidote to messianic notions of the military as savior of the nation from internal enemies.[5] The Greek armed forces appear to be moving toward some sort of accommodation between institutional and occupational features, an accommodation that strengthens military effectiveness and subordinates the military to civic rule.

NOTES

1. For a historical background of the Greek military, see Constantine P. Danopoulos, *Warriors and Politicans in Modern Greece* (Chapel Hill, N.C.: Documentary Publications, 1984). Also see James Brown and Gwynne Dyer, "Greece," in John Keegan, ed., *World Armies*, 2d ed. (Detroit: Gale Research, 1983), pp. 219–228; and Thanos Veremis, "Some Observations on the Greek Military in the Inter-War Period, 1918–1935," *Armed Forces and Society* 4, no. 3 (Spring 1978): 527–541.
2. Dimitrios Smokovitis, "Greek National Defense Policy," *Hellenic Review of International Relations* 3 and 4 (1983–84): 335–380.
3. George A. Kourvetaris, "Professional Self-Images and Political Perspectives in the Greek Military," *American Sociological Review* 36, no. 4 (December 1971): 1041–1057.
4. Dimitrios Smokovitis, "New Professional Trends in the Greek Army: The Five-Year Volunteer Soldiers." Paper presented at the XI World Congress of Sociology, New Delhi, India, August 18–24, 1986.
5. The hypothesis that an element of occupationalism can reduce the praetorian tendencies of armed forces was first presented with regard to civil-military relations in contemporary Italy. Hans E. Radbruch, "Dai valori isituzionali ai valoria occupazaionali: mutamento sociale nell'esercito italiano," *Revista Interdisciplinare Sociologia Militare* 1, no. 1 (1985): 3–35.

XVI

Switzerland

KARL HALTINER

When Charles Moskos first developed the institution-occupation (I/O) thesis, he certainly had the American all-volunteer force (AVF) in mind. By definition, occupational trends presuppose marketlike prerequisites for the structure of the armed forces. Therefore, the armed forces must be organized on a voluntary and professional basis either completely—the AVF—or at least partially—most conscript armies insofar as they feature a professional officer and NCO corps or operate at least part of their forces with volunteers. Can I/O trends also be observed in a pure mobilization model that lacks the professional element and thus the marketlike prerequisites?

An analysis of the development of the Swiss citizen army is well suited for answering this question. Most military and social scientists still consider this army to be the pure form of a militia system. In this chapter, I propose that in spite of the special form of organization in the Swiss armed forces, we can observe changes in the Swiss militia that come very close to those I/O trends described by Moskos.

As far as possible, the following analysis is based on the criteria of the I/O concept. I am not, however, in a position to use some of Moskos's indicators to measure the I/O trends because of the peculiarities of the militia organization of the Swiss armed forces. First, I will outline the militia organization of the Swiss armed forces so that the implications for a possible I/O trend in the militia forces become clear. Then I will examine some I/O trends, first in the normative legitimation of the armed forces and second in their structural development.

THE NORMATIVE BASIS OF THE SWISS MILITIA ARMY

The militia principle is not only the basis for the armed forces; it makes its mark on the political culture of Switzerland, as to federalism and direct

255

political codetermination. The Swiss federal parliament, like all cantonal parliaments, is organized on a militia basis; the federal parliamentarians carry out their legislative duties on an honorary basis in four annual sessions of two weeks' duration. All citizens are called to the ballot boxes several times a year to vote on constitutional matters, bills, or credits on a federal or cantonal level. Thus, the concept of the militia not only refers to the structural pattern of the political and military system, but also implies values and norms for citizens' participation in public affairs.

Because of the special political and social importance of the militia principle, the symbolic functions that citizens' armies generally assume developed comparatively early and strongly in the Swiss military system. They include the role of the military as the "school of the nation," as the symbol of civil honors and of national identity. The degree of military participation has always determined the degree of civil integration and closeness to the social and political centers in Swiss society.

A career in the military requires a certain amount of public service rendered voluntarily and beyond the minimum, but at the same time it assures one's reputation in civilian life. Therefore, a military career is the norm for those civilians who hold leading posts in politics and in the economy. Furthermore, in Swiss society the military assumes the role of an ideal for civic participation. The usual classification of male citizens into "fit for military service," "conditionally fit for military service," and "unfit for military service" is not only a militarily relevant stratification, but also a rather significant social classification. Article 16 of the current military code warns every male citizen of exclusion from personal services if he degrades the army by his way of life.

If the army is questioned in Switzerland, the citizens' model and public symbol is questioned simultaneously. The centuries-old belief in the political importance of the militia principle and the symbolic connection of the military with such values as independence and freedom establish an institutional continuity that as a reference point of collective identity, is significant for the integration of the military into the political and social system of Switzerland.[1]

THE STRUCTURAL BASIS OF THE SWISS MILITIA ARMY

The Absence of a Standing Army

The permanent military institutions in Switzerland are confined, in regard to personnel, to a small corps of instructional officers and NCOs and, in regard to infrastructure and organization, to a framework of basic schools and training centers. The militia services exist in a continuous state of mobilization and demobilization of individual troops throughout the year without a comprehensive basic organization of military personnel. This situation prevents the character of a "total institution" from developing in the militia forces; they maintain a relatively high degree of openness toward their civil environment.

Universality of Military Obligation

All Swiss males are drafted in their nineteenth year. About 10 to 20 percent are disqualified; the rest are accepted for enlistment.[2] Men who are unable to perform any military service for medical reasons pay compensation taxes. Women may serve a shorter term on a voluntary basis.

Long Duration of Service with a Series of Short Training Periods

After undergoing basic training for 17 weeks, the male Swiss citizen must perform another 32 weeks of military service in refresher courses of 1 and 3 weeks over a period of 30 years (35 years for officers). The peculiarity of the militia is that it does not constitute a reserve army: during his service, the militiaman remains an active member of a war formation with all the attendant circumstances (personal equipment at home, armament, continuous training). Thus, the relationship of a male citizen with the armed forces is maintained over a long period. He is under training throughout his prime.

The Absence of a Professional Corps

The most typical feature of the Swiss militia is that aside from the very small corps of professional instructors, all officers and noncommissioned officers have the same status as the troops. Every careerist begins as a soldier and must pass through the ranks by serving as leader and trainer. Thus, the soldier who wishes to rise to a higher rank must attend the proper schools and earn his new rank by serving in a recruit school as a trainer and a commander. In the course of a career, voluntary services add up. A militiaman serves for 330 days, a sergeant for 647 days, a captain about three full years, and a colonel almost five years.

The militia system depends on the extent to which sufficiently talented young men can be called upon. The law permits coercion or compulsory service, if necessary, to secure sufficient leadership staff.

Voluntary Involvement

In their spare time, the commanders of troops on all levels do a large amount of preparation for military service and for the administration of their personnel. According to recent studies, the time spent on military matters beyond actual service amounts to 24 working days a year for a captain and 28 working days for a colonel. These services, like many other voluntary contributions, are crucial to the proper functioning of the militia system.

The Swiss soldier receives only a symbolic payment. Today it amounts to $1.30 (4 francs) a day for a soldier, $1.70 (5 francs) for a noncommissioned officer, $4.70 (14 francs) for a captain, and $7.70 (25 francs) for a two-star general. A salary compensation payment covers part of the income losses caused by the compulsory refresher courses. This payment varies according to the family status of a person, but compensates for at least 50 percent of the lost wages. The payment for longer-lasting voluntary career service compensates only part of the income loss.

Until 1945, the Swiss militia was a mass infantry army, but the more mobile warfare of World War II and developments in arms technology gave rise to serious doubts as to whether it should continue in that form. After long debate, the government and parliament decided in the early 1960s to introduce a new strategic concept, combing static defense with more mobile warfare so that mechanized and airborne operations could supplement a dense infantry protection of the area.[3] This new concept accelerated the organizational and technological development of the militia army. In 1982, the infantry amounted to only 35.7 percent of all troops, whereas the proportion of technical and logistical special troops and technical combat troops (mobile artillery, tank troops, air force) has risen considerably. In regard to armament and organizational structure, the Swiss army hardly differs from other Western European armed forces.

SOME IMPLICATIONS OF THE I/O TREND FOR THE SWISS MILITIA MODEL

Obviously the Swiss militia force contains most of those elements that Moskos considers typical for an institutional type of military organization. Less obviously, however, the effectiveness and preparedness of the militia essentially depend on the dominance of institutional values and norms, notably the voluntary quality of the military career and civilian support for the military.

The voluntary nature of the military career is one of the most important prerequisites for the functioning of the militia system. It ensures that professionalization of the officer and NCO corps will not be necessary in the near future. Coercion is no alternative over the long term: absent or poor motivation among involuntary officers or NCOs would have a disintegrating effect on the armed forces' efficiency. Only the traditional legitimacy of the Swiss militia, which is based on social esteem and high prestige in civil life, offers those noncash benefits that guarantee a high supply of volunteers. This traditional legitimacy allows the volunteers to select a military career that ensures the optimal use of civilian leadership and training in the military system. Thus the readiness to assume leadership voluntarily is an important indicator of the institutional legitimacy of the Swiss armed forces.

Civilian support includes those services that are rendered directly or indirectly to the militia by the society without legal obligation to do so, but simply because of the traditional ties with the military. These services include not only the unpaid work done by the commanders but also the voluntary furnishing of land for shooting practice, the provision of civilian structures such as school premises, offices, and farm buildings for the troops' refresher courses, and the toleration of noise from tanks, planes, and shooting in densely populated central Switzerland. Because civilian sympathy and involvement have been taken for granted, the army has confined itself to a minimal permanent infrastructure to be used exclusively by the military. An increased legitimation of the military according to marketplace standards rather than the institutional criteria of community thinking would lead inevitably to a decline of these predominantly voluntary services. Thus, occupational trends in the militia may be understood as reflections of I/O changes in the normative structures of civilian society.

The normative basis of the militia model is therefore essentially institutional. As shown above, the social rewards for voluntary participation by officers and NCOs, as well as the motivation for performance by militia soldiers, are not based on cash benefits but on the value of civilian honors and the corresponding social prestige for the service rendered. A weakening of those values, a moving away from an institutional format to one that resembled an occupation, would shake the fundamental value base and the organizational efficiency of the militia.

Apart from normative aspects, we must ask whether the impulses for I/O trends do not originate with technological developments, and whether trends in modern armament do not impose an occupational pattern on the militia, whether it is wanted or not. Only insofar as modern weapons and communication systems can be mastered and handled logistically by a militia, the armed forces may maintain political and economic legitimacy in the militia. We must consider the constraints that compel the Swiss militia, like other armed forces, to adapt their organizational structure to new technological developments.

The Willingness to Assume Leadership Functions: 1970–85

Reports from military schools and courses of the early 1970s abound in complaints about difficulties in recruiting NCOs for further training and the declining interest in officers' careers. During the 1970s, roughly a third of all NCOs had risen involuntarily or only conditionally voluntarily to their current rank; the refusals occurred in all branches of the army. Even traditionally renowned arms of the services, such as the tank corps or the artillery, often had high coercion rates.

Verbal judgments in school reports for 1979 showed a trend away from coercion and back toward voluntariness. In the 1980s, because of the economic recession, a military career seems to have regained some attraction as a reference for a civilian career. This trend corresponds to the higher rate of volunteers for the armed forces in all Western nations. A substantial increase in lost-salary compensation for all career services after 1982 might also have helped. In 1972 and 1973, parliament had repeatedly demanded such financial stimuli for a military career. Apparently occupational motives for participation are replacing institutional motives. An analysis of the motives for refusing leadership functions shows a high degree of plausibility for such a change as described by the I/O thesis. Among the reasons for refusing an NCO career listed in military reports between 1972 and 1979, "private" and financial interests—that is, occupational aspects—account for two-thirds of the reasons for refusal. In this motivational structure, we can see a shift of priorities: just as a military career used to be determined by the prospect of public service and self-sacrifice, the new career model orients itself primarily by self-interest and the balance of private benefits.

Changing Sympathy and Involvement

A similar change can be observed in civilian sympathy and involvement with the military. Increased opposition to the extension of military installations and the use of the civilian infrastructure have indicated this trend for years. A large part of the population is no longer willing to accept noise and other emissions caused by the military; resistance (demonstrations, boycotts, and so on) has increased steadily since 1945. The reasons do not seem to be predominantly antimilitaristic; rather, different interests clash because the same territory, central Switzerland, often must serve the economy, tourism, leisure, and the military simultaneously. In this competition, the military is losing its edge.

Changes in the Attitude toward the Militia in Swiss Society: Some Survey Results

Surveys that have been conducted repeatedly since 1976 show a fundamental change in the social valuation of the citizen army. It is manifested in the shift from a *traditional* legitimacy, which is expressed by assigning a positive and central role to the armed forces for Swiss society as a whole, to a more *instrumental* view of the military as a necessary evil. The traditional component of this attitude decreased between 1976 and 1983 in favor of the instrumental[4] (see Table 16-1).

TABLE 16-1. Changes in the Evaluation of the Armed Forces in Switzerland, 1976–83

| | TOTAL | | | AGES 20–29 | | |
	1976	1983	Diff.	1976	1983	Diff.
			(in percent)			
General attitude toward the military						
Traditional (institutional)	49	42	−7	33	24	−9
Instrumental (occupational)	45	52	+7	51	62	+11
Do away with military	6	6	0	16	14	−2
	N = 1,831	N = 1,739			N = 312	N = 341

TABLE 16-2. Institutional and Occupational Aspects in the Attitudes toward the Swiss Armed Forces

| | Attitudes toward armed forces | |
| | Traditional | Instrumental |
	Agreement in percent	Agreement in percent
Attitudes toward military structure		
Institutional aspects		
Military uniform: honor	88	54
Military: citizen duty	90	71
Military: individual efforts for a more efficient army	96	81
Occupational aspects		
Military: more civilianization necessary	39	50
Military: no individual efforts for a more efficient army	21	46

1983 Survey, N = 1,784.

In the terms *traditional* and *instrumental*, the affinity to the I/O typology can easily be observed. In fact, the traditional and instrumental attitudes correlate closely with the criteria for institutional and occupational attitudes, respectively, as noted by Moskos (see Table 16-2).

The traditional attitude stands for values like honor, country, and civil duty and places the efficiency of the military above individual self-interest. The instrumental attitude is determined by individual utilitarian considerations and by the need to organize the military in a more civilian way. The citizen army thus bases its legitimacy less on traditional or institutional loyalty and increasingly on instrumental or occupational factors. The military is no longer the "stronghold of national identity," but is perceived as a national service like the railway or the post office. The awareness of a complementary relationship between a citizen's rights and a soldier's duties tends to weaken, and the attitude toward the military becomes more distant. To a growing degree, it consists of cost-benefit considerations, and thus we observe a loss of institutional legitimation. As symbolic ties are lost, the change in I/O values may also be called disenchantment (*Entzauberung*, as by Max Weber), or secularization.

The loss of institutional legitimacy seems to be one of the main reasons for the decrease in motivation to pursue a military career in Switzerland (Table 16-3). The degree of readiness for a military career is significantly lower for those who view the military from an occupational perspective than for those who regard it from an institutional point of view. For those who orient themselves by traditional values, a military career is a social must; for those who view the armed forces instrumentally, the readiness for a career depends on private cost-benefit considerations. Insofar as occupational legitimacy increases and institutional legitimacy decreases, young people can be expected to avoid entering a voluntary military career in the future.

I/O value trends in the legitimation of the armed forces have consequences elsewhere than in the personnel sector. Sympathy and involvement also decline with the decrease of institutional legitimacy. The critical attitude toward emissions and demands by the militia increases with the occupational point of view; thus, the steady increase in actions against military installations can be explained plausibly. In the long run, the militia might be forced to withdraw behind barrack walls, initiating a process of isolation previously unknown in Switzerland.

CHANGES IN THE STRUCTURE OF THE MILITIA FORCES

To some extent, changes in the normative basis reflect the process of structural differentiation to which the militia army has increasingly been subject since 1960.

The Militia and the Adaptation of Modern Military Technology

Military experts in Switzerland agree unanimously that militia and technology do not contradict each other. Recent experiences show that modern military technology is easier to operate than the older technology and complies with the needs of the militia.[5] The introduction of the computer, for example, allowed the militia to control more sophisticated antiaircraft systems.

Swiss experts have become masters at assessing whether new weapons and technologies may be used successfully by the Swiss army, with its short and interrupted periods of instruction time. An example from the air force may illustrate this point. In 1947, Switzerland bought 75 English Vampire DH-100 jet aircraft. Later the Swiss manufactured 100 slightly modified models of

TABLE 16-3. Professionalization in the Swiss Army: Development of the Number
of Professional Officers and NCOs in Absolute Numbers and by Ratio of Trained Recruits
per Instructor

Year	Draftees	Instruction Officers (Total)	Draftees for Instruction Officer	NCOs as Instructors (Total)	Drafted Men per Instruction NCO
1960	28,214	388	72.71	440	64.12
1965	38,193	458	83.49	600	63.66
1970	34,428	544	63.29	763	45.10
1975	35,095	574	61.14	870	40.33
1980	37,143	638	58.21	912	40.70
1981	38,090	630	60.46	913	41.71

this type under license in Switzerland and adapted them to militia conditions. Today militia pilots fly technically modified versions of the English Hawker Hunter Mk. VI, the French Mirage III S, and the American F-5 Tiger II. (Militia pilots undergo a basic training like that of other militia soldiers, but their initial training takes about one year instead of 17 weeks. Besides their usual refresher courses, militia pilots take individual one-day training flights every month.) After 1986, militia soldiers will operate a Swiss version of the modern German Leopard II tank.

Professionalization

Although the technological development of weapons complies with the militia and does not require professionalization for those who use them, there is considerable pressure for professionalization among the instructors and those who service this weaponry. Although the politically cherished militia ideal and the constitution do not favor tendencies toward professionalization, an increasing professionalization of the militia forces, both latent and manifest, has been observed since World War II.

Manifest Professionalization

The Instruction Corps, which constitutes the small professional framework of the militia army for training the militia officers and NCOs and supervising of the training of the troops, grew steadily between 1960 and 1981, not only in absolute numbers but also in the number of recruits to be trained (see Table 16-3). The number of professional NCOs rose faster than that of officers. The NCO corps primarily includes instructors for troops with a high rate of technology, so the rise in professionalization is obviously a result of the technological development of the army.

Manifest Professionalization: the Air Force and That Observer Corps of Fortresses

Experiences during World War II had shown that the Swiss neutrality could be guaranteed only if the country was protected by permanent air-raid precautions. Therefore, in addition to the regular air force organized on a militia basis, the Swiss army has maintained a "surveillance squadron" with 100 professional pilots and servicemen since 1941. In addition to continuous patrol flights and standby duties, these professionals train the militia combat pilots. The professional observer corps of fortresses also dates back to World War II, when enormous efforts were made to create the so-called Reduit, a series of fortresses to convert the Swiss Alps into a large stronghold. The observer corps maintains and looks after these. It is also responsible for the depots of ammunition, materials, and provisions kept therein, and guarantees the rapid distribution of military resources in case of mobilization.

Latent Professionalization: Military Administration

In addition to this manifest professionalization, a latent professionalization is also taking place, as indicated by the increasing number of civilian employees, usually civil servants. Of the 15,000 employees of the Federal Military Department in 1980 (10,000 in 1950), only 11.2 percent are concerned with actual administrative tasks, 18 percent work as trainers (essentially in the instructional corps), and 5.5 percent procure new armaments. The majority (62.6%), however, maintain materiel and military installations.[6] Arsenals are looked after mainly by civilian personnel who may be militarized in case of mobilization. Military airports and control of the air space, which have been increased systematically over the years, are also operated by civilian technicians. The majority of the military administrative personnel in Switzerland therefore does not concern itself with administrative tasks as such, but provides direct military standby, a job that is done predominantly by members of the armed forces in other nations. Even though the total of professional personnel (19,000) make up only 3 percent of the total militia forces (625,000), which is minimal in comparison with other countries, professionalization constantly rises.

Latent Professionalization at the Top Levels of the Officers Corps

Since 1913, the positions of corps and division commanders have been professionalized, so that militia officers striving for a two- or three-star rank must give up their civilian professions. None have done so since the 1970s; not only the highest ranks of the army but also the command of regiments have been taken over increasingly by members of the professional instruction corps. The reason for this development is not that professional officers of the instruction corps are privileged, but that high-ranking militia officers are less willing to give up their civilian professions. Their attitude becomes understandable when we consider that the rank of colonel may not be attained before the age of 40, an age when civilian achievements do not allow one to make changes easily. Nevertheless, the latent professionalization of the top ranks in the armed forces is regarded in Switzerland as alien to the system.

Opening the Armed Forces to Women

A further international trend to be included in the I/O thesis can be observed in Switzerland as well: the increased opening of the armed forces to women and the process of granting equal rights to women. Although women have been allowed to perform voluntary military service since World War II, when their participation rate was highest, only about 6,000 women are serving in the militia system, and this number is decreasing. An important reason for the low participation rate may be the legal status of women in the armed forces, until recently, women were part of the auxiliary services. This branch of service, restricted to men who are only conditionally fit for medical reasons, has its own ranking system, which differs from the regular one, and generally has low prestige.[7] Since 1945, women have been fighting for equal status with the regular members of the armed forces, and they succeeded in 1985. Although they still perform unarmed service predominantly in the logistical and sanitary branches of the army, they may attain the same officers' and NCOs' ranks as regular members of the army in spite of shorter terms of service. In this sense, women enjoy privileges.

CONCLUSIONS

Initially, I asked whether it was sensible to talk of I/O trends in a militia system that can be compared only conditionally to professional or partly professional armed forces. Accordingly to my analysis, I can answer as follows:

As a concept of analysis, the I/O thesis may also be applied successfully to those armed forces that are neither entirely nor partially organized on a professional basis, provided that the I/O trend in the armed forces is comprehended not only as an intramilitary development but also as a reflection of an I/O modernization process embracing the society as a whole. This modernization is taking place in all the Western industrial nations, including Switzerland.

Because the Swiss militia is relatively permeable to the civilian system and features a permanent organization only as a framework and not as a personnel focus, the I/O trends do not become visible in the inner structure, as they do in other armed forces, but in the *input* area of the military. Insofar as institutional aspects of legitimation, such as the social prestige of the military service, and traditional values of duty, honor, and country lose importance in favor of private utilitarian elements of legitimacy, we note a decrease in the offer of voluntary military participation and of sympathetic involvement, which the militia needs for effectiveness. Compensating for the loss of voluntariness is possible by offering financial stimuli and by resorting to legal coercion, but this step promotes professionalization of the armed forces and thus jeopardizes the militia principle as an organizational pattern.

As the example of Switzerland shows, the increase in occupational elements is not merely a consequence of changes in normative structures, but also of a development in military technology that affects the militia army as much as other types of armed forces. The necessary organization and technical differentiation do not seem feasible without partially sacrificing the citizen-soldier principle and introducing professional elements in a manifest or latent way.

To summarize, the Swiss militia system constitutes a peculiar case in international comparisons: the institutional element dominates sociopolitical integration and legitimacy in an almost ideal way. Yet, it does not constitute an exception because it also reveals elements leading to a development similar to that described in the I/O thesis.

NOTES

1. See the very good description by John McPhee, *The Swiss Army* (New York: La Place de la Concorde Suisse, 1984).
2. E. Wetter, *Schweizer Militarlexikon*, p. 51.
3. See the short description by A. Ernst, "Geschichte der Landesverteidigung", in E. Gruner, ed., *Die Schweiz seit 1945* (Bern: n.p., 1971), pp. 175–201.
4. Survey of 1976, representative random sampling of 1872 Swiss citizens of both sexes, ages 20–80. Survey of 1983, representative random sampling of 1789 Swiss citizens of both sexes, ages 20–80.
5. See H. Wildbolz, "Technisierung und Milizarmee," in H. U. Muller and P. Hauser, eds., *Landesverteidigung in der Zukunft* (Commemorative publication of 175 years of existence of the Officers' Society in Winterthur and its surroundings, Frauenfeld: n.p., 1981).
6. E. Wetter, *Schweizer Militarlexikon*, p. 102.
7. K. Haltiner and R. Meyer, "The Woman and the Army in Switzerland." Paper presented at the 20th Annual IUS National Conference, "The Interdisciplinary Study of Military Institutions," Chicago, October 22–24, 1980.

XVII

Israel

REUVEN GAL

Among the modern Western societies, the Israeli Defense Forces (IDF) probably come closest to a citizens' army model. The Israeli society represents what may be labeled a *warrior society*. With a total military strength of about half a million men and women (in both regular and reserve services) out of a total population of 3 million, every sixth Israeli citizen is a soldier. Every family in this country has at least one representative (most commonly several representatives) serving in the IDF. Approximately one Israeli of every four has lost a relative in one of the Arab-Israeli wars. With six wars over the last three and a half decades, the Israelis are exposed to a full-scale war approximately every six or seven years. The Israeli military is preoccupied with fighting, or anticipates fighting, at all times. Moreover, in the eyes of the average Israeli, the IDF is the only shield between Israeli's existence and its destruction. Therefore, serving in the IDF is considered a primary, almost unquestionable, duty.[1]

Such background characteristics dictate, almost by definition, most institutional and occupational (I/O) modalities as they are reflected in the IDF. My first thesis is that continuous involvement in wars determines, to a large extent, whether a military organization is more institutional or more occupational in its characteristics; my other thesis, however, is that such an assumption must be reexamined along a time dimension. Even during continuous fighting, independent processes change the characteristics of every military organization, the IDF not excluded.

HISTORICAL BACKGROUND

I will start with a brief historical background of the roots of the IDF and its creation. I will skip the 3,000 years of ancient history, which include a rich variety of Jewish warriors (such as Moses, King David, and the fighters of Masada) and limit my review to the last century.

The IDF was formally established on May 26, 1948, just 12 days after the birth of the new state of Israel and in the midst of the War of Independence. Its foundations, however, go back more than 40 years.

In September 1907, a group of 10 young Jewish pioneers, recent immigrants from Russia to Palestine, met secretly in the town of Jaffa with their leader, an intense Russian Jew named Israel Shochat. This group of men and women had been influenced strongly by the socialistic revolutionary movement of Tsarist Russia and had abandoned their religious background as Eastern European Jews upon immigrating to Palestine. Inspired by the decree of Theodor Herzl, the founder of modern Zionism, they were committed to create "a publicly recognized, legally secured home in Palestine for the Jewish people." The young Schochat argued that achieving the Zionist ideal would be impossible unless Jews were prepared to undertake their own self-defense.

The outcome of this secret meeting was the establishment of an armed Jewish organization for the first time in 2,000 years — Hasomer (the "watchman"). Its members, all volunteers of exceptional motivation and quality, became the elite of the small Jewish community in Palestine. In their ideology, they were committed to three major goals: self-defense, the promotion of modern Zionism, and socialism. Before they disbanded during World War I, these founders of the future Israeli armed forces had demonstrated that Jews could be daring fighters, capable of protecting Jewish lives and property.

This historical background is worth mentioning to illuminate the importance of the value system as the underlying core mission of military organizations. I believe that concepts such as duty, honor, and country cannot, by themselves, provide a value system for a military unless they represent real needs. For the young *Hashomer* members such words were, indeed, real and meaningful.

At the end of World War I, Palestine became a British mandate with a mixed Jewish-Arab population. The increasing threats to the lives of the Jewish settlers from their Arab rivals required a drastic solution. The formation of the *Haganah* ("defense") in 1920 was the response of the Jewish community in Palestine to these ongoing threats. This clandestine paramilitary Jewish self-defense organization was the precursor of the IDF.

The labor Zionists who founded and staffed the *Haganah* perceived their martial responsibility as more than protecting the entire community of Palestine Jewry. They viewed it as an inseparable part of a wider ideal of reconstructing Jewish life in the land of Israel, based on humanistic-socialistic principles of justice, righteousness, and social solidarity. Along with their commitment to stand armed against any hostile attacks, they also saw their mission as that of creating "a new society of farmers and workers who would subdue the land by the sweat of their brow."

Thus, from its early days the Haganah was involved in nation-building affairs, including the emigration of thousands of Jewish refugees from pre-

Nazi Europe, the foundation of Jewish settlements at various strategic points, and, in general, providing leadership for the growing Jewish community in Palestine.

During the years that preceded the establishment of the state of Israel and the formation of the IDF, several paramilitary groups other than the *Haganah* evolved: the *Palmach* ("striking companies," composed of young volunteers who were mostly *sabras*, or Israeli-born); the *Stern Gang* (or LEHI) and the *Irgun* (two radical right-wing organizations who considered the *Haganah* too passive); and the Special Night Squads (SNS), initially trained by the legendary Ore Wingate, who later established nonconventional standards of fighting for the *Palmach* and the *Haganah*. All these groups, each characterized by its own strong ideology, molded and shaped the nature of the Israeli military.

The following citation, a brief description of the *Palmach* spirit, is an example of the institutional ideology that can be traced to the contemporary IDF:

> Essentially the Palmach, with about one fifth of its members girls who participated in all actions, constituted a "youth movement in arms", with its own egalitarian style, defiant both of bourgeois values and of external discipline as exemplified by the British Army. . . . Unrecognized, unpaid, often short even of elementary necessities, the Palmach compensated by developing a collective personality of its own. It proudly considered itself more than just a military unit, but a living communal elite, a "fellowship of fighters" . . . Its uniforms, if they can be so called, were khaki shirts and shorts, with the shirt commonly worn outside, supplemented by stocking caps and sweaters. The commanders were young. . . . Rank conferred no privileges. There were no badges of distinction, all lived under the same conditions, ate the same food, and did the same work. The only special right accorded to commanders was that they were expected to lead during an attack and stay behind to cover a retreat—a concert that became embedded in the ethos of the Israeli Army.

Another historical event that affected the nature of the IDF was the Holocaust. The memory of this trauma, in which six million Jews were exterminated without any means of self-defense, is preserved in the mind of every Israeli, including the younger generations who have never been in Europe. This collective memory created a genuine feeling that "this must never happen again" and plays a major role in the legitimacy of and the motivation for military service in Israel.

The official formation of the Israeli army, the IDF, took place during Israel's War of Independence (1947–49). From its very beginning, the IDF's characteristics—in size, equipment, doctrine, and spirit—were dictated by continuous wars and by actual combat needs. The heritage of the *Haganah*, the *Palmach*, the *Irgun*, and the LEHI, and that of its predecessors, the *Hashomer* and the SNS, had many facets. It provided the new IDF with structure (a unified military, based on territorial sovereignty), leadership (a highly trained and dedicated cadre of combat-experienced commanders at all levels), and fighting spirit (a close-knit bond among unit members that emphasized a morale level and a code of conduct as high as their military skills). It also instilled a vital sense of purpose and a deeply rooted democratic loyalty within its members.

From 1948 to the present, the IDF has withstood six wars: the War of Independence, the Sinai Campaign, the Six-Day War, the War of Attrition, the Yom Kippur War, and the Lebanon War. In all but the last, the IDF has had the full support of Israeli society. In fact, Israel's society and Israel's military have always been perceived as indistinguishable.

To understand this perception more clearly, a description of the IDF structure is in order. Israel's armed forces comprise three components or types of military services: A permanent ("neva") service, a relatively small cadre (about 10 percent of the total strength) of career officers and noncommissioned officers; a compulsory ("nova") service composed of drafted conscripts in which the men serve three years and the women two years; and a large body of standing reserves ("milluim"), including all those who have completed their compulsory service.

Modeled partially on the Swiss system of reserve service and partially on the *Haganah's* experience, the Israeli system became exceptional in that the reserve forces were its most important operational components rather than merely an appendage to the regular forces. This combination of compulsory service based on a full draft (about 90 percent of each annual cohort of Israeli males are drafted) and reserve service, which is compulsory for all males up to the age of 55 and all women up to the age of 35, makes military service in Israel indistinguishable from daily life.

This unique quality of Israel's military service was described by David Ben-Gurion, Israel's first prime minister and the founder of the IDF.

> Every Israeli in good health . . . enters adulthood with two or three years of military service. Then he or she goes back to normal life for three-quarters of the time, with the other quarter devoted to reserve training. The result is that at any moment of the day or night the butcher, the baker, the office receptionist, . . . the farmer, the university professor, the shopkeeper, the Israeli man and woman in the street can grab a rifle, or hop into the driver's seat of a tank or behind the complex control panel of a sonar . . . device and be ready to perform his or her military duties with utmost competence. In that sense, Israel's Defence Force functions as . . . the Minute Men of the American Revolution who in seconds could exchange farming implements for rifles in the cause of their country's freedom.[3]

Even the permanent segment of the IDF, that small group of career servicemen and women, lacks the typical career orientation that can be found in many modern military organizations. Rather, it may be characterized, as it was by Ben-Gurion, as exhibiting a "civilian orientation."

> What I mean by that is that our military is very far removed from career-oriented, caste-bound tradition. The IDF is not a club, a lifetime sinecure, a catch-all for people with nothing better to do than dress up in uniform.[4]

With the option of retirement at the age of 40 and with division commanders (holding the rank of brigade general) averaging age 35, the Israeli career officers are generally extremely young.

This means even officers who are serving full time can expect to leave active duty early in life and pursue subsequent civilian careers. It also means we don't have . . . a staff of conformists at the top, perpetually fighting the last war. In our position, we could hardly afford that.[5]

Indeed, these two factors—the reserve system and the early retirement age of IDF officers—have often been cited as two examples of structural mechanisms that have prevented the Israeli Officer Corps from becoming a military caste.

INSTITUTIONAL AND OCCUPATIONAL MODALITIES

With this historical background, I will now turn to a more detailed analysis of the current Israeli army along several of Moskos's I/O modalities.

Source of Legitimacy

The sources of legitimacy for serving in the IDF are rooted in the normative values of Israeli society, which incontrovertibly sustain the full-scale military draft system. Furthermore, throughout the four decades in which this compulsory system has operated in Israel, a stable figure of 99 percent of all eligible conscripts have reported willingly for recruitment on their due day, without any need for threat or warning. Conscientious objectors are very rare, and draft dodging is unheard of.

Among the permanent service members, the basic attitude is quite similar; in most cases the principal motivation for their military career is the call of duty. Though the salaries are relatively high, they are considered less than adequate compensation for the heavy demands and frequent risks involved in their way of life.

The reservists normally are called for 40 to 60 days of active duty a year, with no monetary compensation and without a need for coercion, as long as their service is perceived as critically needed and fully legitimate. We will return presently to this aspect of legitimacy.

Role Commitment

All servicemen and women in the IDF are expected to accomplish any type of mission, at any time, at any place. Because of the nation-building nature of the Israeli army, soldiers of the IDF—conscripts, permanent careerists, and reservists alike—frequently find themselves occupied in nonmilitary missions such as educational programs, treating juvenile delinquents, founding settlements, and supporting distressed communities. This variety of duties makes the role commitment of IDF members very diffused, in Moskos's terms.

Basis of Compensation

During their two or three years of mandatory service, Israeli conscripts are not paid except for some pocket money valued at about $15 a month. Nor do reservists receive any significant payment during their duty periods, though their civilian salaries are guaranteed by a national insurance program. Salaries of the permanent troops are related chiefly to rank and seniority and, as a rule, do not fluctuate according to market needs.

Residence

Unlike the United States military, the IDF maintains complete separation between military bases and civilian housing. With very few exceptions, the IDF does not provide on-post quarters for families; therefore, career officers and NCOs must commute between their duty stations and their homes, which are located in various civilian centers in Israel. Because the workload and operational demands may prevent daily commuting, many families are together only on weekends and not necessarily every week.

In terms of the I/O continuum, this characteristic of the IDF contributes to its institutional nature; the frequent separations between the service person and his or her family make the military into a total institution or a greedy institution.[6]

Societal Regard

Military service in Israel is normally associated with high prestige. During the first years of the IDF, the officer corps represented a true social elite, a cadre of highly talented individuals who were not necessarily interested in military careers but who had answered a call to the most important national demand of that time (indeed, many of them later became Israel's most prominent leaders). Today service in the permanent corps is still highly regarded.

In a survey conducted in late 1974 among a representative sample of Israeli adults, military employment received relatively high scores on an overall career-prestige scale. A colonel in the permanent service was ranked 96 (on a 1-to-100 scale), falling just below a university professor and somewhat higher than a rabbi or a psychologist. A major was ranked 81 (equivalent to a pharmacologist), and a captain (ranked 76) had slightly higher prestige than a commercial airline pilot.[7]

These and other indications reflect the generally favorable attitude of Israel toward its armed forces. This attitude becomes even more positive and appreciative during periods of war or tension, often in the immediate area. Furthermore, this attitude does not necessarily correlate with rank or seniority. In fact, at times Israelis will express even greater gratitude and appreciation to the rank and file than to senior officers.

Post-Service Status

Military service in Israel means more than a legally compulsory duty and a source of status and certain advantages. Military service has become an entrance ticket to Israeli society in general and to the job market in particular. The first thing required of any young person who looks for a job is a certificate of discharge from the military. Not only does it verify that the candidate has fulfilled the requirement for military service, but also it indicates the rank achieved and includes a brief evaluation regarding conduct. In most cases, this certificate is considered far more important than any other form of recommendation.

To summarize the I/O modalities as they apply to the Israeli military, the IDF is evidently an institutional rather than an occupational military organization. In their original form, the Israeli armed forces exempled a calling, an institution dominated by an ideological conviction and a sense of duty. As mentioned earlier, most of the IDF's institutional characteristics stem from the fact that it is a fighting army, a citizens' military, an army that is perceived as a critical shield for the very existence of its country.

TRANSITIONS

No organization endures without changes, and the Israeli military is no exception. Since its formation in 1948, the IDF has undergone some major transitions. I will focus on four such transitions, all of which occurred after the traumatic experience of the Yom Kippur War in 1973.

First, the IDF underwent dramatic growth after 1973: within a 10-year period, it almost doubled its size in manpower, number of units, and quantities of equipment. The permanent corps alone increased from 30,000 to more than 50,000. This huge expansion brought the problems typical to all growing military institutions: lower overall quality of recruits, increased centralization and bureaucratization, increased proportions of non-combatants, and diminished personal contacts. The IDF began to lose some of its unique qualities and began to acquire some of the traits found in other large militaries.

Second, in recent years the IDF has changed gradually from a militocratic model (in which the military is identified totally with the national interest and its legitimacy is unquestioned) to a democratic model (in which the military is legitimate only insofar as its existence and its use of power are supported by societal consent).[8] The results of the last two Israeli-Arab wars exemplify this transition: the 1973 Yom Kippur War, which caught the Israelis by surprise and almost caused the IDF to disintegrate completely, was later labeled *the blunder* and generated enormous criticism of the military. For the first time in Israeli history, a public investigation commission was assigned to investigate the functioning of the military before and during the war. On the basis of the commission's conclusions, several senior officers were forced to resign and many mishaps within the army were exposed and scrutinized. A more recent example is the 1982 Lebanon War and especially the Sabra and Shatila massacre (in which the IDF was found only indirectly responsible), which were followed by wide public criticism. Although most of these criticisms were aimed against the political leaders, the military was also shaken: the IDF changed from a system that had traditionally been almost sacred, beyond any doubt or debate, to a legitimate object of public scrutiny and, at times, critical reservations.

Third, the attitude within the IDF also changed. During the first years of the IDF's existence and after the War of Independence and the foundation of the state of Israel, the military leadership represented a social elite who had responded to a patriotic call. The current officer corps of the IDF, though still highly committed and strongly devoted, is composed primarily of individuals—not always the cream of Israeli youth—who, after completing their compulsory service, decided to pursue a military career and become military professionals. The word *career* is seldom used and the sense of service is still dominant, but the motives for this service have deviated somewhat from the purely ideological. Perhaps the term *pragmatic professionalism*, offered recently by David Segal, best describes the current trends.[9]

Finally, the last few years have seen some cracks in the previously unquestioned base of legitimacy for the IDF. During the recent Lebanon War, and for the first time in IDF history, a senior commander asked to be relieved of his position as brigade commander because, in his opinion, the orders he had been given were "morally unjustified."[10] During the occupation of south Lebanon (1983–85), also for the first time in Israel's history, several reserve officers refused to report for duty in Lebanon on the grounds that they questioned the legitimacy of the IDF's presence there.

All those changes reflect the gradual transformation that the IDF has undergone during the last decade, from a predominantly calling institution to a typical pragmatic profession. Today's IDF represents a specialized, well-integrated body of knowledge and skill and a set of standards and norms that emphasize proficiency and expertise in pursuit of legitimate goals. This, in fact, is Etzioni's definition of a professional body.[11] Thus, the Israeli military is a professional organization that maintains its institutional characteristics, but these characteristics are not as pure and idealistic as they were initially. Societal changes and historical developments have combined to shape the IDF into what it is today.

NOTES

1. For more a detailed description of the IDF and its significance within Israel society, see, for example, Reuven Gal, *A Portrait of the Israeli Soldier* (Westport, Conn.: Greenwood Press, 1986).

2. G. Rothenberg, *The Anatomy of the Israeli Army* (New York: Hippocrene Books, 1979), p. 30.

3. David Ben-Gurion, *Memoirs*. Compiled by Thomas R. Bransten, (New York: The World Publishing Company, 1970), p. 89.

4. Ibid., p. 99.

5. Y. Peri and M. Lissak, "Retired Officers in Israel and the Emergence of a New Elite," in G. Harries-Jenkins and J. Van Doorn, ed., *The Military and the Problem of Legitimacy* (Beverly Hills, Calif.: Sage Publications, 1976).

6. E. Goffman, *Asylums: Essays on the Social Situation of Mental Patients and Other Inmates* (Garden City, Doubleday, 1961); L.A. Coser, *Greedy Institutions: Patterns of Undivided Commitment* (New York: The Free Press, 1974).

7. V. Kraus, "The Perception of Occupational Structure in Israel," *Megamot* 26, no. 3 (1981): 283-294.

8. G. Harries-Jenkins, "Legitimacy and the Problem of Order," in Gwyn Harries-Jenkins and Jacques Van Doorn, eds., *The Military and the Problem of Legitimacy* (Beverly Hills, Calif.: Sage Publications, 1976).

9. D. R. Segal, "Measurement of the Institutional/Occupational Change Thesis." Paper presented at the International Conference on Institutional and Occupational Trends in Military Organizations, U.S. Air Force Academy, Colorado, June 12–15, 1985.

10. R. Gal, "Commitment and Obedience in the Military: An Israeli Case Study," *Armed Forces & Society* 11, no. 4 (1985): 553–564.

11. A. Etzioni, *Modern Organizations* (Englewood Cliffs: Prentice-Hall, 1964).

Part Four

Institution versus Occupation Reconsidered

XVIII

Institution Building in an Occupational World

CHARLES C. MOSKOS AND FRANK R. WOOD

We note two different and competing conceptions of the role of social science and policy, which we call the *engineering* and the *enlightenment* models.[1] From the perspective of the engineering model, the approach most congenial to sponsored research, highly trained methodologists collect quantifiable data and test deductive systems of hypotheses. This work may contribute to new theoretical formulations, but fundamentally the engineering model is one of applied research. The main task is to collect data as rapidly as possible in order to solve specific problems. In the enlightenment model, the main objective of the social sciences is to deepen the policy makers' understanding of social institutions by illuminating critical relationships, not to supply specific answers to particular questions. In a political democracy, the ultimate goal of social research is to enlighten the citizens in their own decision-making processes.

Both good and bad research can occur under either model. Yet, whereas the engineering model is concerned with definitive, preferably quantitative, answers to specific questions in order to make concrete recommendations, the enlightenment model directs attention to fundamental and systemic problems rather than to topical issues of the moment. The institution-versus-occupation (I/O) thesis belongs to the enlightenment type of social research. It seeks to increase sensitivity to how broad military organizational changes affect members' attitudes and commitment, which in turn affect organizational effectiveness. The I/O thesis will not give concrete answers, but it will better inform those who must come up with their own answers.

The value of the comparative approach adopted in this volume is that it allows us to peel away extraneous layers of organizational cosmetics and to reach the core of institutionalism. The research undertaken here points to three basic conditions of institutionalism in armed forces. First, people will accept difficulties and hardships if those in charge are seen to be wholly involved in the system and genuinely concerned about it. Second, there must be a clear vision and articulation of what the institution is all about and how the separate parts relate to the core. The third condition, and the one that subsumes the others, is that members of an institution are primarily value-driven, motivated by factors that contrast with the calculative workings assumed to exist in the marketplace.

Institutional leadership is leadership by deed. The leaders themselves must display devotion to the ultimate goals of the organization, even at the risk of career progress. This devotion entails emphasis of the nation over the military, the armed forces over the branch, and the mission over the career. In the long run, this kind of devotion will count for more than interpersonal skills or inborn leadership traits. In plain words, if attention is diverted to satisfying individual advancement rather than to serving institutional purposes, members of the military will begin to think, "If the boss doesn't care what we're here for, why should I?"

A second major institutional imperative is to understand and communicate how the separate parts relate to the central function. The object is not to oppose or even to slow down specialization, but to make specialists part of the whole. Enhancing generalist identification becomes especially important in the technical military, with its pressures toward identification with civilian counterparts. Ostensibly, the continual rotation of jobs is designed to broaden an officer's experience, but often it renders his appreciation too shallow to see the system as a whole. The issue is how to structure military professionalism so that necessarily specialized personnel are reinvigorated continuously in their institutionalism.

The third and overriding major task is to keep in mind that motivation of members in an institution rests more on values than on calculation, whereas the opposite is true in an occupation. We are not so naive as to believe that pecuniary considerations are absent or even minor in an institution, but we are aware of the findings in the research literature: what we call institutional identification fosters organizational commitment and performance exceeding those of an occupation.[2] The armed services require certain behavior from their members that can never be coterminous with self-interest.

To accent our concept of values as the driving force for institutional members, we may contrast it with the "human resources" school of thought. The model of human resources (a fashionable term in military management circles) considers people as quantifiable entities, akin in some way to material resources. Internalization of norms, in contrast, implies a broad definition of organizational tasks and the intensity with which these tasks are carried out. As Peters and Waterman put it in their study of organizational excellence: "The institutional leader is primarily an expert in the promotion and protection of values. . . . Institutional survival, properly understood, is a matter of maintaining values and distinctive identity."[3]

Each of these basic conditions is easy to state but difficult to achieve in practice. Nevertheless, they serve as bridges between abstract principles and practical policy.

POLICY IMPLICATIONS

The I/O thesis is complex because it deals with different levels of organization. Indeed, failure to remain clear about the level of the discussion accounts for most of the ambiguity in empirical research on institutional and occupational interaction. The I/O thesis must be understood on three levels: micro, macro, and the organizational level that lies between. Although these levels continuously interact with one another, each has different substantive concerns and requires different policy prescriptions.

In addition, each level of analysis has an appropriate methodology. The micro-level focus on the orientation and attitudes of individual members relates best to survey methods. The macro-level approach, which focuses on civil-military relations in the broadest sense, is congenial with social historical studies. The intermediate organizational level, which addresses the structural aspects of groups within the military, is served best by the case-study method and by in-depth interviewing of selected small samples of relevant military persons.

Related to research methodology is the vexing question of whether the I/O concept describes a zero-sum phenomenon or multidimensional processes. When we keep the level of analysis in mind, this problem seems manageable. Broadly speaking, surveys of active-duty military personnel show that although institutional and occupational orientations tend to be related inversely, individuals can score high on both institutionalism and occupationalism (not an infrequent occurrence among career military personnel, according to the data).[4] Macro and organizational analyses, however, point to much more of a zero-sum phenomenon: the more institutional the military, the less occupational it is, and vice versa.[5]

The policy implications of the I/O thesis can be understood according to the level of analysis. Accordingly, we turn to concerns in the contemporary American forces, as reflected in four topical areas: (1) recruitment and retention, (2) the military family, (3) sex roles, and (4) organizational commitment and leadership. We will present an I/O preamble to each topical area and then assess policy implications at micro, macro, and especially organizational levels.

Recruitment and Retention

A comparative overview of armed forces reveals that personnel systems are related best to a twofold concept of military organization: recruitment-oriented versus retention-oriented. A recruitment force, found typically in the ground forces, is relatively labor-intensive and physically demanding, requires short enlistments (or draft periods), and has a large proportion of single members. Conversely, a retention force is more characteristic of navies and especially air forces. Such a force is capital-intensive and skill-demanding, requires long enlistments and does not rely on draftees, and has a large proportion of members who are spouses and parents. Because retention forces contain more technical specialties with civilian counterparts, trends toward occupationalism are more likely to be found there than in recruitment forces.

The air force faces pressures for recruitment, as does the army for retention. In addition, different branches or specialties within the same service may be more or less recruitment- or retention-oriented. Thus, the security forces within the air force, for example, tend to be recruitment-oriented within an otherwise retention-oriented service. When scarce personnel resources are allocated, however, the air force (and usually the navy) give higher priority to retention, and the army (and the marines) place priority on recruitment. That is, a service shows its true nature regarding personnel by devoting most of its attention either to recruitment or to retention.

The truism at the micro level is that most people join the military for a combination of institutional and occupational reasons. Since the end of conscription in 1973, however, the trend has been to offer recruitment incentives with a strong emphasis on occupational motives (though institutional appeals are not absent). Indeed, one of the major contradictions in the sociology of the all-volunteer military is the trauma of occupationally enticed recruits confronting the institutional features of the armed forces, even if these features are less pronounced than in the past.[6]

At the macro level, the major change has not been so much the end of conscription per se as the reliance on an explicitly occupational version of military recruitment and retention. The 1970 Report of the President's Commission on an All-Volunteer Force, which set the pattern for the voluntary military, was inspired primarily by laissez-faire economic thought. For various reasons, the architects of the AVF saw the marketplace as the way to recruit and retain a military force without conscription. Yet international comparisons reveal that even in an all-volunteer military, noneconomic motivation can be maintained despite societal trends toward occupationalism. The key element to keep in the forefront is the notion of the military as service to the country and to show how such service distinguishes military members from the population at large.

The intersection of micro and macro trends has implications for policy at the organizational level. In theory, the American services have a uniform compensation system, but in practice service compensation tends to be differentiated by recruitment·or retention. In the 1980s, the army has received recruitment advantages in the form of generous postservice educational benefits, and the air force and the navy gained more leeway in reenlistment bonuses. The prime incentive for retention forces, however, is the retirement system, and this system has come under question in the 1980s. The reduction of retirement benefits—a strong institutional feature—will aggravate retention problems, especially in the technical branches, where skills with civilian transferability are most common. In addition, by gearing monetary incentives to the market, the armed forces become subject to the vagaries of the economy, over which they have no control.

The distinction between recruitment and retention forces, coupled with an appreciation of the I/O thesis, suggests a grand strategy for military personnel planning. It is time to codify the existing compensation trends that differentiate between recruitment and retention needs. In a world of compensation trade-offs, the technical services can forgo recruitment advantages, and the ground forces can reshape their personnel structure to reduce the career portion, thereby cutting total retirement costs in the long run and moderating political pressures to change the retirement system. A variegated compensation policy might be as follows: retain as much as possible of the traditional retirement program, which is essential for air force and navy retention. Maintain the army's recruitment edge, namely GI Bill types of benefits and short enlistments, to foster a citizen-soldier component in the all-volunteer force. Various other trade-offs can be calculated on the basis of enlistment and reenlistment bonuses and compressed versus decompressed pay scales, but the basic compensation policy outlines are clear. Predictability and consistency in compensation are vital for the technical needs of retention forces and for the citizen-soldiers of recruitment forces.

The Military Family and Spouse

An undisputed finding across all Western military systems is the growing conflict between military demands and family priorities. However, recognizing the conflict is not enough; a more refined analysis must specify how different stages of the family affect military career development. We must be especially alert to the different effects of military-family conflict on husbands and on wives.

At the micro level, the key concern is not to force military members to choose between the military and their families. Members may quickly "divorce" the military rather than their families. At the macro level, we see no signs of reversal in the societal trends toward more women in the labor force and in the cultural trends toward women seeking opportunities independent of the husband's occupational role. In the future, wives will be even less likely than at the present to be adjunct members of the military organization.

Males, especially officers, seem in general to put the military first in the early stages of their career, when they are working hard to make their mark in the military profession. This is not particularly inconvenient for most military people because many are unmarried or without children of school age. Over time, however, the emphasis seems to shift steadily toward the family. For the majority, the decline in career involvement is most evident with the approach of retirement eligibility at the 20-year point. This decline is caused in part by realistic assessment of one's future career in the service and in part by financial considerations. In addition, the officers tend in the later years to give greater weight to family needs, such as children's education, wife's career, and emotional involvement within the family.

The role of military wife is undergoing a transformation, although this is yet to be absorbed fully by the senior command hierarchy. In the comparative studies reported in this volume, the striking discovery is that no military in any other Western country expects as much of the military wife as does that of the United States, this at a time when role commitments to the military community are less forthcoming among American wives than previously. With the increasing number of wives entering the labor force, military spouses no longer have either the time or the inclination to take part in the full panoply of military community life. Military spouses have settled increasingly into occupations, perhaps even careers, in the local area. In addition, a wife's support for her husband's military career is shaped not only by her views on her own role and by her feminist values (or lack thereof), but also by her perceptions of how well (or poorly) her husband's military career is proceeding.

The evidence is persuasive that wives' nonparticipation in the military community has not harmed military institutionalism in other Western countries. This point is worth emphasizing. Another, less obvious consideration, somewhat independent of women entering the labor force, is that wives of junior officers are less likely now than in the past to view wives of senior officers as desirable role models. One way to understand this tendency is to imagine how desirable being a general appears to a lieutenant, compared to how desirable being a general's wife might appear to that lieutenant's wife.

The organizational implications for the family in the American military are not clear-cut, but rather than implicitly discouraging wives of career military personnel from seeking outside employment, a neutral or even a supportive position toward such employment may be better—especially when the children reach school age. Although this idea may appear counterintuitive from an institutional standpoint, we offer the provocative hypothesis that lessening the military obligations of contemporary wives may increase their husbands' institutional commitment. One hallmark of an institution is that it deals with its members as whole persons, taking into consideration their lives outside the workplace.

Sex Roles

Greater reliance on female service members is a clear trend in the American military, reflecting both the increasing numbers of women entering the labor force and the end of the draft, with the resulting expansion of the recruitment pool to include women. In 1988, females constituted 10 percent of all military personnel, in contrast to a token one percent 15 years earlier. The number of military women will almost surely increase in the near and middle future. The core issue for military women is not the combat-exclusion rule but the development of a modus vivendi (or something better) among military life, marriage, and especially parenthood. Cross-national comparisons tell us little about female participation, for no other country (not even Israel) has gone as far as America in expanding opportunities for female service members. Nevertheless, some insight on sex roles can be gained from the I/O perspective.

Statistics tell an important story about marital differences between males and females in the armed forces. In 1985, among males in the lower enlisted grades (E-1 through E-3), 27 percent were married; among male senior noncommissioned officers (E-6 and above), 89 percent were married. Among females, 35 percent of the junior enlisted force were married, compared to 26 percent of the NCO corps. In brief, career males in the armed forces are more likely to be married than first-term males; the opposite is true for females. The data for 1985 also reveal a significant difference in parenthood: among married persons in the NCO corps, only 10 percent of males were childless, contrasted to 70 percent of females. Similar patterns are found among officers. Simply stated, career military women are pressured under the status quo either to be single or, if married, to be childless. The institutional demands of military life on career personnel are much heavier for women than for men.

Personnel policies will have to be overhauled if the role of women is to be institutionalized in the armed forces. Without some major accommodation to the responsibilities of motherhood, women cannot take a significant part in the career military force. One policy implication is readily apparent. Childcare facilities, although admittedly expensive, must be expanded and upgraded if women are to be represented proportionately in the career force. Although many single fathers serve in the American military, parental responsibilities have always fallen and will continue to fall disproportionately on women, whatever their marital status. An extensive childcare system is essential for the effective use of women in the armed forces precisely because the military is a demanding institution. Such a system will also be especially helpful to couples in which both the wife and the husband are in the armed forces, an increasing proportion of the married force. That well-designed childcare programs are becoming a premium benefit in the corporate world for recruiting and retaining skilled female employees is no coincidence.

For military mothers, a second policy recommendation is to create a maternity furlough, with two or three (or perhaps more) years off. Such a furlough would involve a new form of reserve status, in which the woman would have certain priority rights to reenter the active force. At the same time, a mother's time on furlough would not be calculated as active-duty time in grade. This policy would relieve the career-oriented woman from having to make an either-or decision between two valued institutions—the family and the military.

From an organizational standpoint, military men and women differ in another way—the portability of their credentials. Whenever a woman moves into a new assignment, she is tested informally in a manner not typically experienced by men. When a woman's performance is evaluated positively, it is not generalized to other females but is considered an individual accomplishment. If her performance is judged unsuccessful, however, her individual failure is likely to be generalized to all women. In either event, females are tested, repeatedly and needlessly, in ways not confronted by men.

The institutional incorporation of military women is fraught with other complexities. Standard mentoring practices are difficult to apply. If a woman superior mentors a female subordinate, undue attention is focused on both women. If a male mentors a woman, suspicions of sexual favoritism may be raised. The practical answer is that mentoring of females must take place, but in less obtrusive ways than among men. Similarly, drawing attention to sexist acts and statements should as much as possible not be a responsibility of females; rather, enlightened males within the military should take the lead in sanctioning sexist remarks, especially in the informal confines of male groups.

The crux of the matter is that the military organization reflects too well the unclear messages of the society regarding the proper role of women in the armed forces. The combat-exclusion rule is indicative, but not the sole measure of the mixed signals as to whether women should be full members of the mainstream. By way of contrast, we may consider the much more clear-cut message regarding equal opportunity for races. One can imagine a black chief of staff, but seeing a woman in the same role stretches the imagination. Because the military institution does not know exactly where it stands with regard to women, we can fairly say that the incorporation of women into the armed forces is more an issue of organizational climate than a matter of female capabilities. The future promises a continuation of the present ambiguity on sex roles in the armed forces.

Organizational Commitment and Leadership

The effects of institutional and occupational trends on organizational commitment and leadership support much of the research and policy concern with the I/O thesis. Although organizational loyalty and leadership are intertwined, they can be separated analytically for treatment here.

The notion of institutional identification with military organization converges with recent studies on organizational commitment. An authoritative summary of the research literature proposes that employee commitment is more likely to be fostered when an individual is socialized into the organization at early stages of membership rather than at later stages.[7] Recent trends in the American armed forces, however, appear to be moving in the opposite direction. As one result of competition in the labor market, the military has lightened and shortened its socialization (as opposed to skill training) in comparison to times past. Although constraints on off-base living are relaxed for junior personnel, on-base living is often seen as a mark of organizational commitment among career personnel. Concentrating instead on institutional socialization at the initial or entry level might be advisable, after which both institutional and occupational features would be allowed to operate for the career force. In other words, conditions should be such that junior members confront a more restrictive military environment, while career members should have more life-style options. A "tough-early, easy-later" mode of military socialization should replace the present "easy-early, tough-later" tendency.

When we examine the content of military socialization, the issue remains complex, but understanding how the service member can comprehend his or her role in the larger institutional picture might be helpful. This comprehension is twofold: on one side, the service member ought to understand clearly how an assignment fits into the unit mission, which fits in turn into the military mission. On the other side, and perhaps more difficult to achieve, the service member should understand the civic and historical links to the country at large, which is placed in turn within a contemporary strategic context.

On both the military and the civic sides, military people must be given a justification for the utility of the armed forces and an understanding of how military force can be applied in a morally justified manner. Institutional identity may be enhanced by patriotic and military rituals, but rituals are not sufficient in themselves for institution building. They must be accompanied by a civic identification with the nation and by an appreciation of the service member's personal role in the military organization. By contrast, contemporary observers of American men and women in uniform report almost a craving for information on the political and military meaning of military service.[8]

Leadership must affirm altruistic norms at the micro or small-unit level. Socialization by deed is much more powerful than socialization by word. The attribute of leadership that is common across all institutional militaries is based on continuous and personalized interaction with subordinates. Leaders are exemplars of the institution; immediate leaders *are* the institution to their subordinates. Thus, leaders' concern with the professional development of subordinates enhances the institution. The present American system of military leadership does not sufficiently reward training one's replacement. While holding competence constant, promotion criteria ought to favor those leaders who are seen to be the most concerned for group improvement and who are most willing to devote extra time to mentoring subordinates. Peer evaluations (and perhaps, in some instances, subordinates' evaluations) must be given weight in the promotion process.

At the macro level, leaders must be aware of the constraints—political, economic, technological, and cultural—on personnel policies. In articulating institutional needs, however, military leaders must not view themselves as pawns in the grip of larger forces. The responsibility of military leadership is to avoid deterministic explanations of organizational change in the armed forces. More than the leaders of any other groups, military leaders must emphasize to the outside world and to their inner constituency that military service is a value-drive activity that ought not be degraded into occupationalism. To do otherwise is to start down the slippery slope toward professional abdication.

First of all, military leaders must be clear in their own minds as to what is distinctive about military forces, namely, their capacity to make war and their utility for foreign policy, which derives from the war capacity. The military must perform actions—the intentional killing or injuring of other human beings and the mass destruction of property—that are condemned in other contexts. In concert with the top political leaders, military leaders must enshrine the proposition that America, as the greatest free power, has the moral and political obligation to preserve its free institutions and that it must equip itself with a military that can discharge these obligations.

After they have articulated the unique and awesome responsibilities of the military institution, the senior military leaders must be seen as concerned and effective in protecting members' rights and entitlements—not in the sense of aggrandizement for military personnel, but as a way to place personnel policies in an institutional perspective. This is not to say that budgetary trade-offs in personnel policy can be avoided, but that the terms of such trade-offs must take into account professional military judgment. At the executive level, more and more personnel policy is taken into the hands of the Office of the Secretary of Defense and its contract apparatus; at the legislative level, more and more personnel policy alternatives are weighed by budget and accounting offices. These two elements encroach on military professional judgment in exactly those areas that matter most to military members. To the degree that military members perceive senior military leaders as ineffective defenders of an institutional compensation system, creeping occupationalism will appear in the rank and file.

Institutional policies at the organizational level, especially but not exclusively in technical forces, must aim to increase the sense of normative integration. In building institutionalism, the leaders and the led must be socialized continuously to understand how the individual's part supports the corporate whole. Conventionally, this socialization occurs in two basic ways: first, by starting all careers in the institutional heartland, such as basic training for the enlisted troops and commissioning programs for officers; and second, through strategies to reunify the hegemony of the whole through professional military education, notably at the staff and war college levels. These two approaches have their limitations, however; basic training and commissioning programs are variable, have limited normative content, and take place before the centrifugal forces of specialization are operative, and professional military education affects directly only a fraction of the total noncom or officer corps. Moreover, no real evidence exists that professional military education programs, as presently designed, increase holistic or institutional thinking in the career force.

We propose ongoing and informal institutional socialization to fill these gaps. Professional development and institutional knowledge cannot be advanced only in the formal military school system. Indeed, in order to reduce the "ticket-punching" quality of professional military education, not entering it in a military person's official record may even be advisable. Enthusiasm for the organization can be enhanced by continuing education programs involving formal and informal seminars, courses given as temporary duty, realistic reading lists, and programs designed and administered by local commanders. A continuing education program would be less selective and more voluntary than standard professional military education. Those who choose to take part in such continuing education—a program that would be much more comprehensive than the present system of professional military education—would increase the number of cadre able to communicate, by practice and by precept, an institutional understanding of the armed forces.

The content of professional military education must be rethought as much as the format. Professional military education aims currently to prepare a military person for promotion, but it should also be seen as a way of broadening a member's experience in the armed services. Professional development should be consistent with articulating a vision of the organizational whole. This point is particularly important because of the degree to which the military departs from the most common form of institutional promotion: being promoted with one's cohort. Because few individuals will make general (and not many will make full colonel, for that matter), exposure to adjacent skills is good not only for a military person's individual morale, but also for increasing the knowledge of how and why the officer does his or her own job.

AN I/O STANDARD OPERATING PROCEDURE

The topical and comparative studies presented in this volume were intended to uncover the fundamental institutional features of armed forces in selected Western countries, beneath all the cultural differences and organizational variations. By placing the American case in a comparative perspective, we have highlighted the dynamics of institutional and occupational trends in armed forces.

Our survey of armed forces in democratic societies charts the following I/O landscape. Institutions are primarily value-driven entities, and occupations are calculative enterprises. A society needs both types of organizations, depending on what need is being served. Real organizations blend both normative and utilitarian features, but effective armed forces must be predominantly institutional because they require a commitment that cannot be bought. Civilian and uniformed leaders must understand and articulate this quality. Furthermore, institutional features can occur in a variety of organizational forms and are not connected intrinsically with either conscripted or all-volunteer forces, though occupational trends are probably more likely in the latter case. Service members who do not display certain traits traditionally seen as institutional should not be read out of the institution. Institutional identity and behavior do not revolve around spending time at the officers' club, spouse's participation in the military community, or on-base residence, though these activities may be consequential at certain times and places. Institutional identity is not necessarily antithetical to specialization and advanced technology. Specialization fragments only when it is accompanied by horizontal identification with civilian counterparts.

The interaction of institutional and occupational trends is not deterministic, but portends a wide variety of potential outcomes. Building institutionalism does not mean that all aspects of occupationalism must be discarded; neither is it necessary to treat the I/O thesis as producing detailed policies of change to cover all contingencies. When ways to reinvigorate institutionalism are found, they will grow from an understanding of how organizational policies shape the behavior of military members, which in turn affects military effectiveness. To gain leverage against policies that foster exaggerated occupationalism, military leaders in particular need a place to stand. This is what the I/O thesis provides.

We must encourage a general attitude, a cast of mind, an outlook rather than a checklist of things to do and not to do. When we are confronted with personnel issues that have I/O consequences, the following procedure should be standard practice. Before we ask what we are going to do or how we are going to do it, we must ask, "What's the story?"[9] How did these concerns develop? Do they foster institutional or occupational tendencies in the military? Start the story as far back as it properly goes. This is especially important in military staffs, where turnover is a fact of life. Plot key trends while entering key events. Locate "old hands"—earlier staffers in the office, Congressional staffers, Department of Defense civilian staff, in-house historians, academics, contract consultants—to track down earlier events, personalities, who is grinding which axe, and the politics of the whole matter.

This simple procedure focuses initial attention on the underlying organizational issue instead of on the question of what to do (which thus is kept at bay a little longer). Going back to the beginning almost always provides a better bearing on institutional implications and helps build resistance to quick fixes. The military profession, more than others, must foster thinking about future unintended consequences of present policies. Elements of both the occupational and the institutional can be recombined in different ways, but some recombinations are more effective than others. What works well for career members of the air force will probably not work as well for army recruits and vice versa. The lodestar, however, is always articulation and enhancement of value motivation.

In sum, a restoration of institutionalism does not imply turning back the clock. Rather, it entails the establishment of a new balance after a long period of indiscriminate acceptance of the marketplace mentality. The implications of the I/O thesis for policy are not to put things back the way they were but to bring about a more satisfactory state of affairs. To look beyond the present requires a clear sense of the past; one must consider what might occur if day-to-day decisions are made without regard to long-term consequences. An institutional sense includes seeing what has the potential to cause undesired changes of direction. By looking back, we look ahead, identifying what is worth preserving from the past and carrying into the future.

NOTES

1. A fuller explication of engineering versus enlightenment is given in Morris Janowitz, "Sociological Models and Social Policy," in Janowitz, *Political Conflict* (Chicago: Quadrangle, 1970), pp. 243–259. A more explicit application of these models to military sociology is Morris Janowitz and Charles C. Moskos, "Consequences of Social Science Research on the U.S. Military," unpublished paper, July 1986.

2. See Richard T. Mowday, Lyman W. Porter, and Richard M. Steers, *Employee-Organization Linkages* (New York: Academic Press, 1982); Amitai Etzioni, *A Comparative Analysis of Complex Organizations* (New York: Free Press, 1975); Rosabeth Moss Kanter, *The Change Masters* (New York: Simon and Schuster, 1983).

3. Thomas J. Peters and Robert H. Waterman, Jr., *In Search of Excellence* (New York: Warner Books, 1982), p. 281.

4. David R. Segal, "Measuring the Institutional/Occupational Change Thesis," *Armed Forces & Society* 12, no. 3 (Spring 1986): 351–375; Michael J. Stahl, T. Roger Manley, and Charles W. McNichols, "Operationalizing the Moskos Institution-Occupation Model," *Journal of Applied Psychology* 63 (1978): 422–427; Michael J. Stahl, Charles W. McNichols, and T. Roger Manley, "An Empirical Examination of the Moskos Institution/Occupation Model," *Armed Forces & Society* 6 (Winter 1980): 257–269; and Michael J. Stahl, Charles W. McNichols, and T. Roger Manley, "A Longitudinal Test of the Moskos Institution-Occupation Model," *Journal of Political and Military Sociology* 9 (Spring 1981): 43–47; Michael J. O'Connell, "The Impact of Institutional and Occupational Values on Air Force Officer Career Intent." Research Report, Air War College, Maxwell Air Force Base, Ala., March, 1984.

5. However, surveys that use especially constructed questionnaire items have found much more of a zero-sum I/O relationships than those reported above. See Charles A. Cotton, "Institutional and Occupational Values in Canada's Army," *Armed Forces & Society* 8 (Fall 1981): 99–110; and Graham J. Pratt, "Institution, Occupation, and Collectivism Amongst Australian Army Officers," *Journal of Political and Military Sociology* 14, no. 2 (Fall 1986): 291–302.

6. John D. Blair and Robert L. Phillips, "Job Satisfaction Among Youth in Military and Civilian Work Settings," *Armed Forces & Society* 9 (Summer 1983): 555–568.

7. Mowday, Porter, and Steers, *Employee-Organization*, pp. 61–62.

8. Charles C. Moskos, "The All-Volunteer Force," in Morris Janowitz and Stephen D. Westbrook, eds., *The Political Education of Soldiers* (Beverly Hills, Calif.: Sage, 1983), pp. 307–322.

9. This mode of analysis is adapted from Richard E. Neustadt and Ernest R. May, *Thinking in Time: The Uses of History for Decision Makers* (New York: Free Press, 1986).

APPENDIX

I/O Questionnaire Items

In the below sets of questions, an institutional (I) or occupational (O) response is based on agreement with the item.

I. Items used by air force researchers; selected from preexisting survey items; principal investigators: Michael J. Stahl, Roger T. Manley, and Charles W. McNichols.

1. If I left the Air Force tomorrow, I think it would be very difficult to get a job in private industry with pay, benefits, duties, and responsibilities comparable with those of my present job. (O)
2. An Air Force base is a desirable place to live. (I)
3. The Air Force requires me to participate in too many activities that are not related to my job. (O)
4. Air Force members should take more interest in mission accomplishment and less interest in their personal concerns. (I)
5. I wish that more Air Force members had a genuine concern for national security. (I)
6. What is your opinion of discipline in today's Air Force? (I = too lenient)
7. More supervision of member performance and behavior is needed at lower levels within the Air Force. (I)
8. An individual can get more of an even break in civilian life than in the Air Force. (O)

II. Items used by Army researchers; selected from preexisting survey items; principal investigators David R. Segal and John D. Blair.

1. In thinking about the kind of job you would like to have, how important are each of the following? A job that
   a. gives me a chance to serve my country well. (I)
   b. gives me a chance to make the world a better place. (I)
   c. is steady; no chance of being laid off. (O)
   d. where the pay is good. (O)
   e. where the fringe benefits are good. (O)

2. Indicate how important were the following in your decision to enlist or accept a commission in the Army:
   a. wanted to serve my country. (I)
   b. to continue a family tradition of military service. (I)
   c. job opportunities looked better than in civilian life. (O)
   d. a secure job with promotions and favorable retirement benefits. (O)

III. Items used in study of Canadian military; items constructed to measure I/O orientations; principal investigator Charles A. Cotton. First three deal with organizational primacy, second three deal with organizational scope.
   1. No one should be compelled to take a posting he or she does not want. (O)
   2. Military personnel should perform their operational duties regardless of the personal and family consequences. (I)
   3. Personal interests and wishes must take second place to operational requirements for military personnel. (I)
   4. What a member of the armed forces does in his or her off-duty hours is none of the military's business. (O)
   5. Differences in rank should not be important after duty hours. (O)
   6. What a member does in his private life should be of no concern of his supervisor or commander. (O)

IV. Items used in study of Australian military; items constructed or borrowed to measure I/O orientation; principal investigators Graham J. Pratt and Nicholas A. Jans.
   1. The service requires me to participate in too many activities that are not part of my job. (O)
   2. No serviceman should be compelled to take a position he does not want. (O)
   3. For an officer, his service should come first and personal matters second. (I)
   4. What a serviceman does in his off-duty hours is none of his service's business. (O)
   5. Service personnel should perform their duties regardless of the personal and family consequences. (I)
   6. My service requires me to participate in many activities that are not part of my job. (O)
   7. Personal interest and wishes must take second place to operational requirements for Servicemen. (O)
   8. Members of my Service should take more interest in getting the job done and less interest in their personal concerns. (I)
   9. An appeals system for promotion is required throughout the Defense Force. (O)
   10. Servicemen need a nonindustrial association (no industrial action) to represent their views on pay matters. (O)

11. Servicemen need a legal advocate to represent them before a pay tribunal. (O)
12. As long as it did not interfere with discipline, a representative association would be a good idea. (O)
13. I am opposed to the idea of a representative association negotiating on my behalf. (I)
14. To me, unionism and the military just do not go together. (I)
15. Even if it did not behave like a civilian union, I still could not support the idea of a serviceman's association. (I)
16. The economic interests of Defense Force personnel are best served by the existing chain of command. (I)
17. The workforce at large has the right to strike and so should servicemen. (O)
18. In this day and age, service personnel need a civilian-style union to defend their interests. (O)
19. I am opposed to the idea of industrial democracy in the services. (I)

V. Other items proposed for future surveys.
  1. Compensation should be based primarily on one's merit and not on rank and seniority. (O)
  2. Compensation should be based primarily on one's technical skill level and not on rank and seniority. (O)
  3. Bonuses and off-scale pay should be directed toward military specialties where there are manpower shortages. (O)
  4. Military members with specialties that require advanced training or a high level of technical skill should be paid more than their counterparts of the same rank. (O)
  5. I normally think of myself as a specialist working for the military rather than as a military officer. (O)
  6. Holding all economic considerations to the side, I would prefer to live in military housing. (I)
  7. Military personnel should be able to live off or on base as they prefer. (O)
  8. The wife of a military officer ought feel as much a part of the military community as her husband. (I)
  9. I would prefer that the dollar value of military "benefits" be added to my pay and the "benefits" dropped. (O)
  10. Military personnel who commit crimes off duty and off post should be tried by a military court martial rather than civilian courts. (I)
  11. If I suddenly became rich (due to an inheritance, lottery winning, etc.), I would continue my military career until retirement. (I)
  12. Service members need some kind of an association (not a union) to represent their views on compensation matters. (O)
  13. The compensation interests of service members are being adequately served by the senior military command. (I)

14. As long as it does not interfere with good order and discipline, military personnel need a union to defend their interests. (O)
15. In today's technical armed forces, we really don't need so much military ritual and tradition as in times past. (O)

BIBLIOGRAPHY

Selected Bibliography on the Institution/Occupation Thesis

Bennett, J. F. (1985) "Professional Attitudes of Canadian Forces Junior Officers." Air War College Research Report, Maxwell Air Force Base, Ala.

Boëne, B. (1984) "The Moskos and Thomas Models Contrasted." *International Forum 4.* Munich: SOWI, 35–66.

Boëne, B. (1987) "Banalisation des armées: le cas français." *Futuribles,* no. 111, 39–54.

Burelli, D. F., and Segal, D. R. (1982) "Definitions of Mission Among Novice Marine Corps Officers." *Journal of Political and Military Sociology,* 10, Fall, 299–306.

Caforio, G. (1987) "Charles C. Moskos." *Sociologia e Forze Armate.* Lucca: Maria Pacini Fazzi, 109–24.

Cotton, C. A. (1979) *Military Attitudes and Values of the Army in Canada,* Report 79-5. Toronto: Canadian Forces Personnel Applied Research Unit.

Cotton, C. A. (1981) "Institutional and Occupational Values in Canada's Army." *Armed Forces & Society,* Vol. 34, No. 8, Fall, 99–110.

Cotton, C. A. (1983) "Institution Building in the All-Volunteer Force." *Air University Review,* Vol. 34, No. 6, September–October, 38–49.

Faris, J. H. (1981) "The Military Occupational Environment and the All-Volunteer Force" in A. R. Millett and A. F. Trupp, eds., *Manning the American Armed Forces.* Columbus: Mershon Center of the Ohio State University, 31–41.

Foote, S. D. (1987) "Building Institutionalism in the United States Air Force," Research Report 87-0850. Maxwell Air Force Base, Ala.: Air Command and Staff College.

Janowitz, M. (1977) "From Institutional to Occupational: The Need for Conceptual Clarity." *Armed Forces & Society,* Vol. 4, Fall, 51–54.

Johns, J. H. (1984) *Cohesion in the U.S. Military.* Washington, D.C.: National Defense University Press.

Johnson, C. B. (1986) "Society's Occupationalism and Its Affect on the Professional Development of Junior Marine Officers." Unpublished doctoral dissertation, Department of Sociology, Northwestern University, Evanston, Ill.

Margiotta, F. D. (1978) "The Changing World of the American Military," in F. D. Margiotta, ed., *The Changing World of the American Military.* Boulder, Colo.: Westview Press, 423–49.

Margiotta, F. D. (1983) "Changing Military Manpower Realities," in F. D. Margiotta, J. Brown, and M. J. Collins, eds., *Changing U.S. Military Manpower Realities.* Boulder, Colo.: Westview Press, 7–38.

Moskos, C. C. (1977) "The All-Volunteer Military: Calling, Profession, or Occupation? *Parameters,* Vol. 7, No. 1, 41–50.

Moskos, C. C. (1977) "From Institution to Occupation: Trends in Military Organization." *Armed Forces & Society,* Vol. 4, 41–50.

Moskos, C. C. (1979) "Les Transformations du metier militaire dans l'armée des Etats-Unis: de la vocation a l'empoi." *Revue des Sciences Politiques,* January–March, 45–56.

Moskos, C. C. (1982) "Institution versus Occupation: Gegensaetzliche Modelle Militaerischer Sozialisation," in R. Zoll, ed., *Sicherheit und Militaer.* Munich: Westdeutscher Verlag, 199–211.

Moskos, C. C., and Faris, J. H. (1982) "Beyond the Marketplace: National Service and the AVF," in A. J. Goodpaster, L. H. Elliott, and J. A. Hovey, Jr., eds., *Toward a Consensus on Military Service.* New York: Pergamon Press, 131–51.

Moskos, C. C. (1984) "La nueva organización militar: institucional, occupaciónal o plural?" *Iztapalapa: Revista de Ciencias Sociales y Humanidades,* Año 5, Nos. 10–11, January–December, 297–306.

Moskos, C. C. (1986) "Institutional/Occupational Trends in Armed Forces: an Update." *Armed Forces & Society,* Vol. 12, No. 3, Spring, 377–82.

Moskos, C. C. (1987) "La banalisation de l'institution militaire." *Futuribles,* No. 111, June, 27–38.

Nuciari, M. (1984) "Institución vs. occupación: discusión y tentativa de adaptacion del modelo I/O a las fuerzas militares italianas." *Iztapalapa: Revista de Ciencias Sociales y Humanidades,* Año 5, Nos. 10–11, January–December, 75–80.

O'Connell, M. J. (1984) "The Impact of Institutional and Occupational Values on Air Force Officer Career Intent." Research report submitted to Air War College, March.

Pinch, F. C. (1982) "Military Manpower and Social Change: Assessing the Institutional Fit." *Armed Forces & Society,* Vol. 8, Summer, 575–600.

Radbruch, H. E. (1985) "Dai valori istituzionali ai valori occupazionali—mutamento sociale nell-esercito Italiano." *FOR.ARM.ES.,*

No. 1, January–April, 1–35.

Roehrkasse, R. C. (1986) "Air Force Enlisted Grade Management." Air War College Research Report No. AU-AWC-86-182, May.

Segal, D. R. (1983) "From Political to Industrial Citizenship," in M. Janowitz and S. D. Wesbrook, eds., *The Political Education of Soldiers*. Beverly Hills, Calif.: Sage Publications, 285–306.

Segal, D. R., Blair, J. D. Lengermann, J., and Thompson, R. (1983) "Institutional and Occupational Values in the United States Military," in F. D. Margiotta, J. Brown, and M. J. Collins, eds., *Changing U.S. Military Manpower Realities*. Boulder, Colo.: Westview Press, 107–27.

Segal, D. R., and Young, H. Y. (1984). "Institutional and Occupational Models of the Army in the Career Force: Implications for Definition of Mission and Perceptions of Combat Readiness." *Journal of Political and Military Sociology*, Vol. 12, Fall, 243–56.

Segal, D. R. (1986) "Measuring the Institutional/Occupational Change Thesis." *Armed Forces & Society*, Vol. 12, No. 3, Spring, 351–76.

Stahl, M. J., Manley, T. R., and McNichols, C. W. (1978) "Operationalizing the Moskos Institution-Occupation Model: An Application of Gouldner's Cosmopolitan-Local Research." *Journal of Applied Psychology*, Vol. 63, 422–27.

Stahl, M. J., McNichols, C. W., and Manley, T. R. (1980) "An Empirical Examination of the Moskos Institution-Occupation Model: Independence, Demographic Differences and Associated Attitudes." *Armed Forces & Society*, Vol. 6, Winter, 257–69.

Stahl, M. J., McNichols, C. W., and Manley, T. R. (1981) "A Longitudinal Test of the Moskos Institution-Occupation Model: A Three-Year Increase in Occupational Scores." *Journal of Political and Military Sociology*, Vol. 9, Spring, 43–47.

United States General Accounting Office (1986) *Military Compensation: Key Concepts and Issues*. Washington, D.C.: Author.

Wood, F. R. (1980) "Air Force Junior Officers: Changing Prestige and Civilianization." *Armed Forces & Society*, Vol. 6, Spring, 483–506.

Wood, F. R. (1982) "U.S. Air Force Junior Officers: Changing Professional Identity and Commitment." Unpublished doctoral dissertation, Department of Sociology, Northwestern University, published as Defense Technical Information Center Technical Report AD A 119101.

Contributors

Bernard Boëne, D.G.E.R., ESM de Saint-Cyr, 56381, France.

John S. Butler, Department of Sociology, University of Texas, Austin, Texas 78712, USA.

William H. Clover, Department of Behavioral Sciences and Leadership, U.S. Air Force Academy, Colorado Springs, Colorado 80840, USA.

Charles A. Cotton, School of Business, Queen's University, Kingston, Ontario K7L 3N6, Canada.

Cathy Downes, Strategic and Defense Studies Center, Australian National University, GPO Box 4, Canberra ACT 2601, Australia.

John H. Faris, Cole and Weber, 308 Occidental Ave. South, Seattle, Washington 98104, USA.

Bernhard Fleckenstein, Social Science Institute of the Bundeswehr (SOWI), D-8000 Munchen 40, Winzererstr. 52, Federal Republic of Germany.

Reuven Gal, Israel Institute of Military Studies, PO Box 97, Zikhon Ya'Akov 30900, Israel.

Karl Haltiner, Institute for Sociology, University of Bern, Speichergass 29, 3011 Bern, Switzerland.

Nicholas A. Jans, School of Administrative Studies, Canberra College of Advanced Education, PO Box 1, Belcone ACT 2616, Australia.

Thomas H. McCloy, Department of Behavioral Sciences and Leadership, U.S. Air Force Academy, Colorado Springs, Colorado 80840, USA.

Charles C. Moskos, Department of Sociology, Northwestern University, Evanston, Illinois, 60201 USA.

Mady Weschler Segal, Department of Sociology, University of Maryland, College Park, Maryland 20742, USA.

Patricia M. Shields, Department of Political Science, Southwest Texas State University, San Marcos, Texas 78666, USA.

Dimitrios Smokovitis, 43 Marathonodromou, Psychikon, Athens 15452, Greece.

Jan S. van der Meulen, Foundation Society and Armed Forces (BMK),

Bachmanstraat 1, 2596 Ja Hague, Netherlands.

Frank R. Wood, National Defense University, Fort McNair, Washington, D.C. 20319 USA.

INDEX

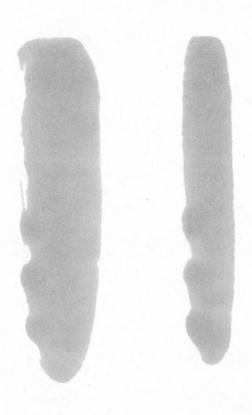

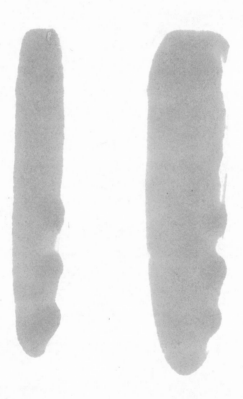